Hazard Modeling and Assessment of the Nile Delta Coast

Kamal Srogy Darwish

Hazard Modeling and Assessment of the Nile Delta Coast

Remote Sensing and GIS Applications

Springer

Kamal Srogy Darwish
Department of Geography
Minia University
Al Minya, Egypt

ISBN 978-3-031-44323-7 ISBN 978-3-031-44324-4 (eBook)
https://doi.org/10.1007/978-3-031-44324-4

This Springer imprint is published by the registered company Springer Nature Switzerland AG
The registered company address is: Gewerbestrasse 11, 6330 Cham, Switzerland

ذَٰلِكَ فَضْلُ ٱللَّهِ يُؤْتِيهِ مَن يَشَآءُ ۚ وَٱللَّهُ ذُو ٱلْفَضْلِ ٱلْعَظِيمِ ۝

This book is dedicated to my parents, my sisters, my brothers, my wife (Hala), my daughters (Aseel and Kenzy), and my sons (Yousef and Mohamed) for their inspiration, encouragement, and help.

Preface

COASTS is considered one of the most morphodynamic landforms on the earth's surface. It represents the transitional area between land and sea. Millions of people live along or near coastlines around the world. *Low elevation coastal zones* (LECZs) around the world are considered vulnerable land to various natural and anthropogenic hazards, including sea level rise, coastal erosion, flooding, and shoreline retreat. The Nile Delta Coast (NDC) is similar to other deltas around the world; it is experiencing shoreline changes due to a combination of natural and anthropogenic driving factors. The NDC has suffered from severe coastal erosion processes and shoreline retreat since the construction of the *Awan High Dam* (AHD) in 1964. It prevented the sediment supply to the coast along the Rosetta and Damietta promontories. Previous global and regional studies have indicated that the Nile Delta is one of the most vulnerable deltas to the risk of sea level rise and coastal inundation in the near future due to climate change, global warming, land subsidence, and increased negative anthropogenic impacts. The NDC is the highly populated coastal zone in Egypt and has a large contribution to the Egyptian economy. This book intends to assess the multihazard risk assessment along the NDC based on a multicriteria analysis approach. Satellite remote sensing (RS)-based digital image processing and geographic information systems (GIS) have recently contributed to analyzing, assessing, and modeling many complicated problems in coastal environments. These geospatial technologies provide reliable and accurate information and solutions that can be helpful for experts and decision-makers to make suitable decisions for sustainable planning and development goals. The satellite-based assessment of coastal changes is a useful technique for obtaining accurate and up-to-date information of shoreline change positions over-time, mapping erosion-accretion aspects, evaluating coastal defense, predicting future shoreline positions, mapping vulnerable areas, and managing coastal resources. Geographic information systems (GIS) have great capabilities to handle, manipulate, and analyze massive amounts of spatial data and information, such as smart systems, and can be employed in modeling the spatial dimensions of the real world. Accordingly, GIS can be used to identify and quantify coastal impacts and risk assessments. The integration of earth observation and geoinformation systems is an extremely helpful technique for historical shoreline change analysis along the NDC since 1945 and for predicting future trends. This book consists of five chapters: Chap. 1, Provides an **Introduction** include location of study area and a meta-analysis of the Literature Review of the study area. The literature review was performed using machine learning of GIS-based modeling of sea level rise, shoreline change, coastal vulnerability, coastal erosion hazard, and disaster risk assessment. Chap. 2, Presents the **Physical Characteristics of the Nile Delta's Coastal Plain**. In this chapter, four different physical characteristics were analyzed, starting with the geological setting, which studied the formation and evolution of the Nile Delta, stratigraphy, geographical distribution of surface deposits, and geological structure. The second setting is the morphology analysis of the study area, including topographic analysis, slope mapping, and aspects. The third setting is the climate of the northern Nile Delta, which includes main elements such as temperature, wind, rainfall, sunshine, humidity, and evaporation. The last setting is the geomorphological and morphometric characteristics of the Nile Delta coast.

Chap. 3, Provides an **Assessment of the Nile Delta's Coastline Dynamics: A Remote Sensing and GIS-Based Computational Approach**. In this chapter, spatiotemporal changes in the Nile Delta coastline between 1945 and 2015 were analyzed using satellite remote sensing data and GIS techniques. A vector-based spatial computational analysis of the multiple historic shoreline positions digitized from the Egyptian topographical maps (scale 25,000) surveyed in 1945 and the automatically extracted shorelines from the multitemporal Landsat imagery were taken in 1972, 1984, 2001, and 2015. The NDWI/MNDWI spectral indexes were applied using the ERDAS Imagine v.2014 modeler to radiometrically enhance the water bodies from the land. The spatial shoreline rate-of-change statistics were calculated using the digital shoreline analysis system (DSAS) produced by the USGS. The ArcGIS-DSAS geodatabase was created and includes the required computational vector layers: shoreline, baseline, and transect layers. DSAS has four spatial computational parameters, which are used to calculate shoreline changes: the net shoreline movement (NSM), shoreline change envelope (SCE), end point rate (EPR), and linear regression rate (LRR). All rate-of-change results have been spatially mapped and analyzed.

Chap. 4, Presents **GIS-Based Spatial Modeling of Potential Impacts of Sea Level Rise Along the Nile Delta Coast**. There are several natural phenomena, including climate change, ocean thermal expansion, and glacial melting from Greenland and Antarctica, that typically cause changes in sea level. The global average sea level is expected to rise through the twenty-first century, according to the IPCC projections between 0.18 and 0.59 cm. NDCs are vulnerable due to the impact of climate change and related long- and short-term sea level rise. The NDC is a low-elevation coast because most of the coastal plain elevation in the Nile Delta region is already under sea level. Egypt is ranked as one of the top five countries expected to be affected by a 1 m sea level rise resulting from global warming. According to this scenario, the total urban areas, Egypt's GDP, natural resources such as coastal zones, water resources, water quality, agricultural land, livestock, and fisheries may be subjected to vulnerability. The methodology in this chapter used ArcGIS v.10.2 Model builder to build a spatial model to simulate and assess the potential impacts of coastal flooding due to SLR and land subsidence along the NDC. Three different stages of the spatial model required a variety of geospatial data. (1) The topographic dataset was built from digital elevation models (DEMs), which were generated from space-based SRTM v.1 data. (2) Land-use/land-cover map, which was classified using maximum likelihood supervised classification from Multispectral Landsat 8 OLI/TIRES. (3) Population density map over the coastal administrative governorates along the NDC (Alexandria, Behira, Kafr El Sheikh, Damietta, and Port Said).

Chap. 5, Provides a **Geospatial Assessment and Mapping of Coastal Erosion Vulnerability Along the Nile Delta Coast: A Remote Sensing and GIS Approach**. In this chapter, beach erosion risk has been studied and mapped along the NDC. The objectives of this chapter are to (1) map the erosion and sedimentation hazards along the Nile Delta coast, mainly tackling the risk on beaches, harbors, archaeological and heritage sites, and promontories. (2) Evaluation of coastal protection engineering along the Nile Delta coast (seawalls, jetties, breakwaters, and groins). Moreover, (3) a GIS-based vulnerability assessment of NDC using CVI-based multi-criteria decision analysis includes eleven physical and coastal processes and marine climate factors (geomorphology, geology, slopes, elevation, land subsidence, annual shoreline advance/retreat rate, coastal protection, relative sea level rise waves, tides, and population density). In general, this book has various methods and techniques of geospatial technology applications, which are used to assess coastal hazards and risks in one of the most populated/vulnerable coasts in the world. Satellite remote sensing and GIS-spatial modeling approaches were successfully utilized to map hazardous coastal areas and suggest mitigation methods for every type of hazard. The contribution of this book may be helpful and useful for spatial planning experts, GIS specialists, Geomatic engineers, research in the field of coastal remote sensing, geographic information systems, coastal zone management using geoinformatics, geomorphology, oceanography, and coastal engineering. The publication of this book would not have been possible without help, support, and positive motivation from my great professors *Scot E. Smith* (Florida, USA) and *Magdy Torab* (Alexandria, Egypt). Great thanks to Mr. Arumugam Deivasigamani from Springer Verlag for his help, patience, and motivation.

Finally, I would like to thank my family, wife *Dr. Hala Abdelmalek*, and my four adorable kids, *Aseel, Yousef, Kinzy, and Mohamed*, for their wonderful help, support, and patience in allowing me to spend time to complete this book.

Al Minya, Egypt
June 2023

Kamal Srogy Darwish

Contents

List of Figures

List of Tables

Chapter 1
Introduction

Abstract This chapter aims to discuss the concepts of coastal hazards and disasters, geomatics technologies such as remote sensing and geographic information systems and their applications in coastal and marine hazard modeling, mapping, and assessment along the Egyptian Nile Delta coast. This chapter is divided into three different and essential parts. The first part focuses on geospatial technology and future trends in coastal hazard mapping, modeling, and assessment. The second part of this chapter presents the location of the study area, and the last part uses machine-learning techniques to analyze the literature reviews.

Keywords Nile Delta coast · Coastal hazards · Remote sensing · Geospatial modeling

1.1 Overview

The Nile Delta is considered one of the oldest world delta systems in the world. Herodotus in 450 B. C was the first to describe its triangular shape as "Delta" since it resembles the inverted Greek letter Δ. Geologically, the formation and evolution of the Nile Delta started between the upper Miocene and present by the sedimentary processes and the alluvium brought by the old seven active branches of the Nile (Stanley and Warne 1993).

The ancient Nile Delta distributaries completely disappeared due to siltation processes and were replaced by the present Damietta and Rosetta branches. Geomorphologically, most of the Nile Delta Coast (NDC) is a muddy-sand beach, with coastal flats distinctive from sand accumulations and sand dunes, shallow lagoons, headlands, sabkhas, wetlands, and fish farms. The northern coastal zone of the Nile Delta is a low-elevation coastal zone with the exception of small sand dune features along the Burullus headland. Therefore, it is considered a vulnerable zone to long-term sea level rise and extreme climate impacts. The NDC is one of the highly populated coastal zones in Africa and Egypt. It is ranked as one of the most populated coasts in the world; it is occupied by more than 20 million people. In addition, it has many

K. S. Darwish, *Hazard Modeling and Assessment of the Nile Delta Coast*,
https://doi.org/10.1007/978-3-031-44324-4_1

large cities, villages, roads, fertile agricultural lands, tourism resorts, archeological and geoheritages sites, harbors, and investment projects. The dams constructed on the river Nile channel during the last century have contributed to reducing the fluvial sediment supply, which fed the coast. As a result, severe coastal erosion risk zones and shoreline retreat appeared along NDC promontories and headlands that increased dramatically after the construction of the Aswan High Dam (AHD) in 1964. Egyptian governments have built a series of coastal protection devices to mitigate NDC from marine and coastal hazards, such as erosion risk, sea level rise, and storm surges. On the other hand, the tectonic structure of Northern Africa has affected the NDC and caused land subsidence of the coastal plain, which has increased the potential impacts of SLR and coastal flooding along the coast. Previous studies indicate that rising sea level within 2 m will lead to shoreline retreat at a distance of approximately 60–80 km southward from the current position (Hereher 2010).

This book is dedicated to multihazard risk modeling and assessment that threatens the NDC, including coastline dynamics, sea level rise, land subsidence, tsunamis, and severe coastal erosion. In recent years, several previous studies have introduced field investigations, such as archeological, sedimentological, geomorphological, and geological indicators of ancient marine and coastal hazards and disasters along the NDC, which caused substantial damage, destruction, loss of ancient cities and civilization, and submergence of ports during the Holocene. Currently, geospatial information systems and technologies have helped research multisource data capturing, storage and arrangement, accuracy assessment, and spatiotemporal analysis. In this book, the application of satellite remote sensing and geographic information systems is mainly used for coastal hazard assessment and modeling along the NDC.

1.1.1 Satellite-Remote Sensing

Remote sensing (RS) refers to the technology of acquiring information about the Earth's surface (land/ocean) and the atmosphere using sensors onboard airborne (aircraft and balloons) or spaceborne (satellites and spacecraft) platforms (Wang 2010). RS is a comprehensive Earth observation (OE) technology that collects, processes, and provides an image of electromagnetic wave information (radiated/reflected) by long-range targets using various sensing instruments without directly contacting objects to detect and recognize various objects on the ground. Each object on the surface of the Earth has specific electromagnetic wave reflection or radiation properties, and information about these properties reaches the remote sensing sensor through the atmosphere. The sensor creates a distant sensing image by measuring how strongly an object on the ground reflects an electromagnetic pulse. Therefore, remote sensing imagery is essentially a record of electromagnetic radiation and ground object interactions. Electromagnetic waves and their basic characteristics are the basis of understanding the principle of remote sensing imaging (Li 2020). Remote sensing has been identified as one of the primary data sources to produce land-cover maps that indicate landscape patterns and human development processes (Darwish and Smith 2023).

Satellite imagery has greatly contributed to the mapping of coastal ecosystems, changes, and coastal zone management. Landsat remote sensing data have been globally used in coastal applications for several decades (Ritchie et al. 1990). The capabilities of multispectral data have helped to monitor and measure the biophysical characteristics of coastal habitats (Colwell 1983). Moreover, the multitemporal data of these satellites were used to track changes in the characteristics of coastal environments (Wang and Moskovits 2001). Advances in satellite sensor design and data analysis techniques make remote sensing systems more practical and desirable for use in coastal ecosystem management, such as estuaries, coastline dynamics and forecasting, wetlands, and coral reefs. Satellites provide a wide spatial coverage, consistent revisits, and multispectral images for the analysis of coastal environmental transformation with different spatial resolutions. Recently, a variety of multispectral and hyperspectral sensors/images have contributed to the highly accurate monitoring of coastal land-cover changes, shoreline mapping, water quality assessment, and mapping of biotic/abiotic suspended particle concentrations. High-resolution satellite imagery is now available for a variety of coastal applications, such as coastline delineation, land-use/land-cover change analyses, wetland mapping, and oil slick tracking.

Very high spatial resolution remote sensing images and digital elevation models (DEMs) are widely used in coastal management applications (Klemas 2015). Very high spatial resolution imagery (few centimeters) can be obtained from the photogrammetry-based UAV drone system, which is a useful technique for coastal applications, mapping land cover, and planning. UAVs are powered aircraft operated remotely or autonomously with preprogrammed flight planning. The two main types of UAV configurations are fixed wing (airplane) and rotary wing (helicopter), which are now being used effectively in coastal environmental studies (Klemas 2015). Although multispectral medium spatial resolution remote sensing data such as Landsat, SPOT, ASTER, and Sentinel images have been broadly applied to solve a variety of coastal problems, the complexity of coastal constituents imposes significant challenges in the application of remote sensing technologies due to the dynamic spatial and temporal nature of coastal habitats. Recent developments in active remote sensing, including LiDAR (light detection and ranging) and interferometric synthetic aperture radar (InSAR) technologies, as well as optical remote sensing, such as hyperspectral and high spatial resolution sensors, have brought new types of data and enhanced capacities for coastal environment studies.

LiDAR and InSAR have been effectively used for coastal habitat mapping, morphological change analysis, topographic deformation, and identification of ancient submerged/buried landscapes and archeological features. Moreover, hyperspectral remote sensing employs hundreds of narrow and contiguous spectral bandwidths in data collection to enhance the capacity of coastal habitat identification and mapping. Finally, RS technology has a large contribution to coastal landscape and seascape mapping. It has been globally utilized for assessing and mapping the dynamics of coastal landscapes and land-cover/land-use changes caused by natural and anthropogenic forces (Darwish and Smith 2021; Torab and Azab 2006).

1.1.2 Geographic Information Systems (GISs)

Geographic information systems (GIS) are defined as computerized systems aimed at capturing, storing, querying, analyzing, and displaying geospatial data (Chang 2014). It is also defined as a specialized computer database program designed to collect, store, process, retrieve, and analyze spatial data (Steinberg and Steinberg 2015). It can help users collect, process, edit, store, manage, share, analyze, model, and visualize large volumes of datasets to understand spatial relationships, patterns, and trends and make educated and sound decisions (Tian 2017). (GIS) have been widely used for decades in coastal zone management, monitoring dynamic changes and evaluating the damage caused by coastal hazards such as storm surges, hurricanes, and tsunamis, in combination with remote sensing data.

Coastal data infrastructure (SDI) has become very important for coastal zone management and disaster prevention because of its ability to combine multisource geospatial data in geodatabases. In the era of geoinformatics and geospatial technologies, the integration of geographic information systems, remote sensing, and ground surveying can be valuable techniques for assessing and mapping spatiotemporal coastline dynamics, mapping vulnerable coastal erosion zones, evaluating coastal engineering protection devices, and modeling the potential impacts of sea level rise. Many useful spatial computational tools working in the GIS environment include the digital shoreline analysis system (DSAS), which has the ability to quantitatively analyze historical and future locations of shorelines based on multitemporal shoreline positions, baselines, coastal protection systems and urbanized databases, geology, and the marine climate. GIS has a huge set of spatial analysis functions and tools that can help integrate and correlate different types of data (*vector, raster, tabular, and descriptive*) to create specialized geodatabases for spatial analysis and modeling. The integration of remote sensing, GIS, and multicriteria decision-making analysis (MCDA) used for assessing the coastal vulnerability index (CVI), which uses a set of significant variables/criteria, can be georeferenced. In addition, GIS has the ability to predict the future impacts of coastal and marine hazards, and it is used effectively for coastal engineering protection work evaluation (Darwish et al. 2017).

1.2 Study Area

The Nile Delta Coast (NDC) is located along the Egyptian Mediterranean coast in northeastern Africa, as shown in Fig. 1.1. It extends along Egypt's Nile Delta coast from Alexandria in the west to the El Tina coastal plain (Northern Sinai coast) in

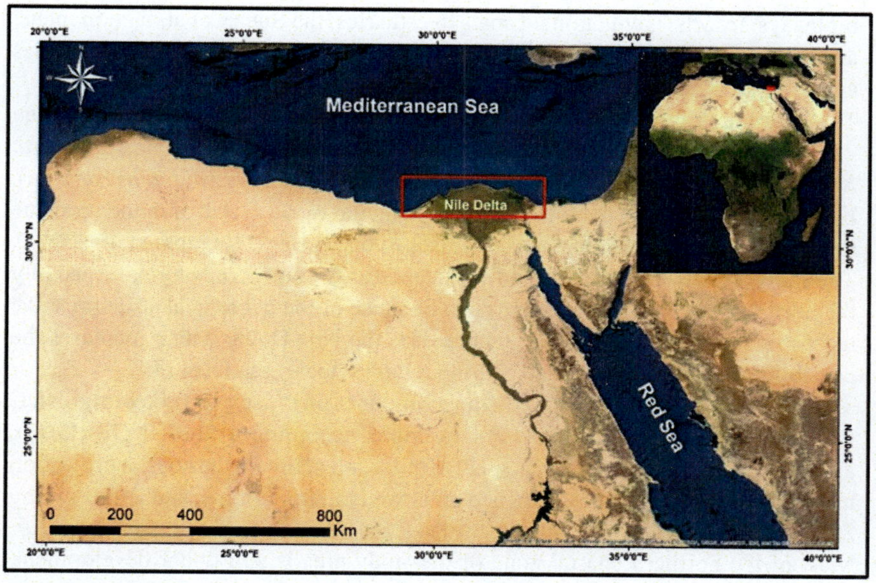

Fig. 1.1 Location of the Nile Delta coast

the east for a distance of 370 km. The study area is located between longitudes 29° 19′ 56″ E–32° 38′ 22″ E and latitudes 30° 39′ 06″ N–31° 36′ 20″ N.

The NDC and its coastal flats are partially backed by coastal flats, dunes, or brackish lagoons. The coastal plain consists of five shallow coastal lagoons from west to east: Marriott, Idku, Burullus, Manzala, and Port-fouad, all of which are connected to the sea through narrow artificial inlets called (bugaz). This coast is distinctive with two large protruding promontories called Rosetta and Damietta, separated by embayments and saddles. The formation and evolution of the Nile Delta are related to seven large ancient branches that built the present delta; they are gradually silted up and replaced by the present-day Rosetta and Damietta. Many populated coastal cities and villages distributed along the NDC include Alexandria, Rashid, Kafr El Sheikh, Baltim, Damietta, and Port Said.

There are several reasons for the importance of this study:

- The NDC is suffering from a significant coastal change because of the lack of fluvial supply to the Nile discharge and sediment load, where the amount of sediment that was feeding the mouths of the Nile diminished, reaching near zero after the start of building the high dam in 1964 (Frihy and Dewidar 2003; Abd-El Monsef et al. 2015). Previous studies have confirmed that the Nile Delta coast has been exposed to severe coastal erosion and shoreline retreat since the beginning of the nineteenth century in some coastal zones of the NDC (Frihy et al. 2003). It increased dramatically after the construction of the Aswan High Dam (AHD) due to the reduction in sediments that fed the coast (Ghoneim et al. 2015). Coastal erosion poses a threat to the Nile River promontories, drains, beaches, coastal highways, resorts, navigational canals to estuaries, the Suez Canal, lagoon inlets, archaeological sites, and ports, as well as threatening the shores of coastal cities and villages (Darwish et al. 2017).

- The study area is one of the most populated areas in Egypt; it has more than 25 million people distributed over nine coastal governorates (*Alexandria, Behira, Kafr El Sheikh, Gharbia, Sharqia, Dakahlia, Damietta, Port Said, and Ismailia*). Egypt is ranked among the top five countries in the world at risk of inundation due to coastal flooding when the sea level rises to one meter (Dasgupta et al. 2009).
- The study area is vulnerable to land subsidence on the northern margins, especially the northwestern northeastern margins. Because of the geological structures, the most important is the hinge line that crosses the Nile Delta, which increases the possibility of subsidence because of the relative (local) sea level rise.
- The Egyptian government spends huge amounts of money to protect the NDC from coastal erosion processes. For example, according to the UNDP (2009), approximately $16.8 million was spent to build and maintain the coastal protection means of concrete seawalls, detached breakwaters, short groins, and jetties during 2009–2014 along the NDC.
- Most of the northern coastal zone of the study area is still under development. Because of the problems of coastal erosion and SLR, the intrusion of salty water and its mixing with groundwater increases the salinity and composts the soil and groundwater. The NDC has valuable mineral deposits that arrived to the coast from the river Nile suspended sediments. The beaches of the study area also contain many mineral resources, the most important of which are the black sand beaches that are distributed along the Burullus area and are exposed to loss by coastal erosion.

1.3 Literature Review

In this section, a systematic review of the literature mainly used remote sensing and geographic information system (GIS) techniques for assessing coastal erosion hazards, shoreline change, and sea level rise along the NDC.

Machine generated summaries

Hazard Modeling and Assessment of the Nile Delta Coast

Machine generated keywords: slr, headland, plain, impact slr, index, climate change, climate, northern, ras bar, dem, vulnerability, canal, beach

1.3.1 Section 1 (Sea Level Rise, Climate Change, Coastal Vulnerability)

Machine generated keywords: slr, impact slr, index, dem, canal, satellite, parameter, pollution, climate change, value, climate, oil, vulnerability, soil

---

GIS-Based Approach to Estimate Sea Level Rise Impacts on the Damietta Coast, Egypt

https://doi.org/10.1007/s12517-021-06810-3

Abstract-Summary

Damietta Governorate is in the eastern part of the Nile Delta, and it is likely to face serious impacts due to SLR because of its low-lying sandy coast. This study investigates the actual impacts of SLR and associated shoreline retreat on the Damietta coast using a geographic information system (GIS) despite current uncertainties about SLR. Shoreline retreat at the Damietta coast was estimated using the Bruun rule. In a 2-m SLR scenario, the submerged area will expand to 661 km^2, which represents 64% of Damietta's area. Upon estimation of the annual shoreline retreat, this was predicted to range from 2.7 to 3.8 m/year at the Damietta coast under the impact of SLR. SLR thus poses a serious threat to the Damietta Governorate as its land is subsiding at a high rate, which is a major factor that amplifies the severity of potential inundation. There is a serious need to protect Damietta's coast and its human population from the impacts of SLR.

Introduction

In terms of its topography, the delta is mainly flat, and one-third of its land contains areas that lie below sea level and are less than 1 m (Hereher 2010), making the impacts of SLR extremely serious there. SLR could have critical social and economic impacts on the Nile Delta, including loss of agricultural land and aquaculture resources, damage to infrastructure, population migration, salt intrusion, and salinization of groundwater resources (Elsharkawy et al. 2009). Damietta Governorate is in a low-lying coastal zone, where SLR can have serious impacts. The Damietta Governorate occupies an area of approximately 920 km^2, which represents 4.6% of the Nile Delta. The total area of the delta governorates is as much as 2879 km^2, of which 631.74 km^2 is built-up land (Abdrabo and Hassaan 2015); Damietta represents 32% of the delta governorates' area and 6.8% of its urban area.

Methodology

The most recent land-use map dates back to 2010, which was used for overlaying with the inundation zones to identify the impacts of SLR. The most common approach used to model the inundation area is the bathtub approach, where the area is inundated if its elevation is less than or equal to the projected sea level. Inland areas that are lower than or equal to the projected sea level may unintentionally be mapped as inundated because they are not hydrologically connected to water bodies (Oswald and Treat 2013). An alternative approach developed by Poulter and Halpin (2008) considers that the area will be inundated if its elevation is less than or equal to the projected sea level and is directly connected to seawater. The GDEM data were reclassified based on the SLR scenarios (i.e., 1 m, 2 m), where the elevation of any area below the corresponding SLR scenario was considered to result in inundation.

Results

Under the 2-m SLR scenario, the submerged area will expand to 661 km^2, which represents 64% of Damietta's area. These considerable losses in the coastal zone actually occur with a eustatic rise in sea level of only 24.4 cm, which represents between 24.4 and 63% of the total RSLR in the 1-m and 2-m scenarios. The total length of railways in Damietta is approximately 45.5 km, but only approximately 350 m of railways would potentially be affected in the 1-m scenario. With a potential 1-m increase in SLR, the agricultural canals and drainage systems would lose 70 km of their length, which represents 6.2% of the total canal length. For a 1-m rise in sea level and up to the next 100 years, the shoreline erosion (R) related to the predicted sea level rise could be in the range from 2.69 to 3.79 m/year at Damietta's coast.

Discussion

Beach erosion in Damietta Governorate is controlled by four major factors: first, the accelerated SLR; second, the shortage of riverine sediment input due to dams along the Nile River; third, the land lowering due to tectonic and sediment compaction over time; and finally, the anthropogenic activities along the coast, which are either activities for SLR adaptation (e.g., beach nourishment, jetties, groins) or naturally occurring events, such as erosion of the coastal sand bar separating Manzala Lake from the sea (Hereher 2014). In his study, Hereher (2010) concluded that one-quarter of the Nile Delta region, including Damietta Governorate, would be inundated if the sea level rises by only 1 m. However, the inundated area in this research was an underestimate because the rate of land lowering was not taken into consideration.

Conclusion and recommendations

The major effect of SLR on the Damietta Governorate is land subsidence, which is at the highest rate in the eastern area of the Nile Delta, where this governorate is situated. By 2100, the land will have subsided by 0.75 m, increasing the impact of SLR. Several essential sectors, such as residential, transportation, agricultural, industrial, and tourism, as well as settlements along the coast of Damietta and further inland, are under serious threat from SLR. Environmental awareness of the impacts of SLR among individuals could play an effective role in adapting to and reducing the costs of SLR.

Climate Change and Its Impacts on the Coastal Zone of the Nile Delta, Egypt
https://doi.org/10.1007/s12665-009-0135-0

Abstract-Summary

The main objectives of the current work are (1) to determine the historic pattern of shoreline changes (erosion and accretion) along the north coast of the Nile Delta, (2) to present a future view on what to be expected regarding climate change impacts, sea level rise (SLR) scenarios, expected land losses and alteration of some soil characteristics, (3) to recognize the negative impacts of SLR on the Nile Delta coast, and (4) to assess and suggest protection measures. The investigated area with 394 measured locations is located along the northern coast of the Nile Delta between Alexandria and the ElTina plain in the Sinai Peninsula between 29° 20′ and 32° 40′ E and 29° 54′ and 31° 35′ N with a minimum erosion value of 1.11 m^2, a maximum of 6,044,951.64 m^2 and a total of 16.02 km^2. A total of 177 sites showed minimum accretion values of 0.05 m^2, maximum values of 2,876,855.86 m^2, and total values of 13.19 km^2. Mediterranean SLR along the Nile Delta coast could be estimated considering three different scenarios (low 0.20 m, medium 0.50 m, and high 0.90 m). Over the coming decades, the Nile Delta will face greater threats due to SLR and land subsidence as well. Regarding climate change and its impacts on soil characteristics, a rapid increase in salinity values during the first three decades was found. Some protection measures were assessed and suggested to combat or tackle SLR.

Introduction

One of the most certain consequences of global climate change is accelerated global sea level rise (SLR), which will intensify the stress on many coastal zones, particularly where human pressure has already diminished natural and socioeconomic adaptive capacities. The coastal wetland ecosystems of Egypt, such as salt marshes, are particularly vulnerable to rising sea levels because they are generally within a few centimeters above sea level (IPCC CZMS 1992; IPCC 2007). The Nile Delta is highly threatened not only by SLR but also by land subsidence.

Data collection and analyses

Historical data belonging to some soil characteristics dated to 1978 were obtained from MASR (1978), while those pertaining to 2001 and 2008 and the results of the current work, respectively. Digital processing of Landsat 5.0 TM and Landsat 7.0 ETM + satellite images dated to 1980 and 2003 using ENVI 4.5 software (ITT 2008). Geographic information system works were executed using ArcGIS 9.3 software (ESRI 2007). The following activities were performed: the ArcMap editor function was used to digitize coastlines on both 1983 and 2003.

Results and discussion

The first topic deals with determining the pattern of shoreline change along the north coast of the Nile Delta (erosion and accretion processes), the second topic is to present a future view on what is expected concerning SLR scenarios and climate change effects on soil properties, the third topic is the impacts of SLR on the coastal zone, and the fourth topic is an assessment and suggestion of some protection measures. Without the aforementioned sand belt, the water quality in coastal freshwater canals will be altered, groundwater will be salinated, recreational areas and beach facilities will be inundated, and 6,900.47 square kilometers of cropland, wetland, and fish ponds representing 28.93% of the total area of the Nile Delta will disappear. The adjacent agricultural areas will alter salt-affected soils due to high capillary forces in these heavy clay soils. Organic matter and salinity were detected as the variables most affected by climate change in the coastal area of the Nile Delta.

Conclusion

The impacts of climate change will be very serious, as one-third of Egypt's fish catches are made in or near the northern lagoons. Recreational tourism beach facilities would be endangered, and essential groundwater would be salinated. This would cause serious groundwater salination, and the impact of increasing wave action would be serious. Special natural and scenic characteristics of the coastal area are being damaged by erosion and ill-planned development. Some soil characteristics, such as salinity and organic matter content, have been altered, and the process is expected to increase in the future.

Recommendations

Because global warming may result in substantial SLR with serious adverse effects in the coastal zone, coastal governorates (Port Said, Damietta, Dakahlia, and ElBehaira) should anticipate and plan for such an occurrence. In light of competing demands and the urgent need to protect and give high priority to natural systems in the coastal zone, present state and local institutional arrangements for planning and regulating land and water uses in such areas are inadequate. The key to more effective protection and use of the land and water resources of the coastal zone is to encourage the coastal governorates to exercise their full authority over the lands and waters. Local governorates should be assisted in cooperation with the central government in Cairo in developing land and water use programs for the coastal zone, including unified policies, criteria, standards, methods, and processes for dealing with land and water use decisions of more than local significance.

Vulnerability Risk Assessment and Adaptation to Climate Change-Induced Sea Level Rise Along the Mediterranean Coast of Egypt
https://doi.org/10.1007/s11027-012-9418-y

Abstract-Summary

The consequence of sea level rise (SLR) on the Mediterranean coastal areas in Egypt, particularly the Nile River Delta, has become an issue of major concern to Egypt's population and the government. Previous publications disregard the entire Mediterranean coast of Egypt as an integral unit subject to the impacts of the SLR. This study aims to analyze the risks, rank the vulnerability and suggest adaptation measures to mitigate the impact of SLR along the Mediterranean coast of Egypt. The social and biophysical vulnerabilities demonstrate the asymmetrical impacts of SLR on the Mediterranean coast of Egypt.

Introduction

Significant changes in climate and their major impacts, such as sea level rise (SLR), are already visible globally. Out of its belief in the consequences of climate change and its potential impact, Egypt, together with the international community—since the Earth Summit held in Rio de Janeiro, Brazil, in 1992 and the development of the United Nations Framework Convention on Climate Change (UNFCCC)—has taken part in most of the studies and research activities, conferences, and meetings addressing this phenomenon. This is consequent to the absence of national climate change and SLR predictive scenarios, evaluation of the effectiveness of the existing protection measures, and the boundaries of the likely impacts. Its primary aim is to analyze risks, to rank vulnerability, and to suggest adaptation measures to mitigate the impact of SLR along the study area. The study objectives are to identify the potential risks of SLR impacts along Egypt's Mediterranean coastal area.

Methodology

Limitations for the present study to undertake full quantitative climate change risk assessment for the Mediterranean coast of Egypt are as follows: scarcity and quality of data, absence of national long-term climate change predictive scenarios, uncertainties and assumptions, difficulties in measuring potential loss, and probability of occurrence. Considering the above, qualitative risk analysis is employed. The following is the process employed to conduct the vulnerability-risk assessment: identify and characterize the impacts of SLR on the entire Mediterranean coastal area of Egypt, assess social and biophysical vulnerabilities and ranking, assess current and future risk and identify the main risks, propose adaptation measures in the absence of long-term national predictive climate change and SLR scenarios, and consider the IPCC (2007) assumptions. This study follows Egypt's National Strategy for Adaptation to Climate Change and Disaster Risk Reduction (IDSC/UNDP 2011).

The Mediterranean coast of Egypt: physical setting

It is characterized by a series of distinct geomorphologic features, including beaches, carbonate ridges, inland depressions (low-lying areas), and harbors. Some low-lying areas include the beaches, Lake Maryut, the southeastern depression neighboring, and the area at El Tarh. The broad beaches of the Delta are partially backed by dunes, cultivated land, urban centers, and coastal wetlands forming natural lagoons and reclaimed fish farms. Three brackish coastal lagoons (Idku, Burullus, and Manzala) are separated from the sea by sandy barriers breached by artificial inlets. In this sector, the coastline from the Tineh plain to El Arish is distinguished by the existence of coastal sand dunes, 10–15 km wide, which rise up to approximately 25 m above MSL. The low-lying areas are the beaches, Bardawil lagoon, and the Tineh plain.

Relative sea level has been estimated from tide gauges installed at Alexandria, Abu Qir, Rosetta, Burullus, Damietta, and Port Said (Frihy et al. 2010b).

Vulnerability assessment and impacts of sea level rise

Oil and gas exploration and production is a rapidly growing activity in the Egyptian Mediterranean coastal areas, particularly in the Nile Delta. The loss of coastal wetlands will have consequences for many sectors, such as food production, water treatment and nutrient cycling functions, and wildlife habitats (Nicholls et al. 1999). Although 80% of Egypt's agricultural land is on the Nile Delta and old valley, agriculture in the northern delta coastal area has a small contribution to the total production and economy because of the low land quality (grade) of this sector. Engineering structures (revetments, jetties, groins, detached breakwaters, and seawalls) established by the Egyptian Shore Protection Authority "ESPA" along the Mediterranean coast of Egypt have multifaceted purposes: a) protection against erosion and b) control of beach changes and subsequent sedimentation at the harbor's entrances, river estuaries, drain mouths, and lagoon inlets.

Vulnerability ranking

The vulnerability of the Mediterranean coast of Egypt to the impact of SLR is ranked as low, moderate, and high. The consolidated Pleistocene carbonate ridges provide natural protection against the impacts of SLR and storms. At Alexandria, the shoreline associated with the natural consolidated carbonate ridge (67% of the total coast) is substantially longer compared with those armored by engineering structures (~20%) and the unprotected ones (~13%). It is characterized by a relatively low relief of + 1 m above MSL, a high rate of subsidence, and rapid erosion due to the combined effect of SLR and coastal processes.

Qualitative risk assessment

Findings including the definition of the problem, vulnerability assessment and ranking, and qualitative risk assessment are employed to identify potential risks and to display the protection means that have already been utilized. An unpredictable number out of the 10 million inhabitants in the coastal areas of both the northern Nile Delta and Alexandria will be at risk, particularly those who reside in the identified highly vulnerable areas. The economy will be in danger since several activities rely on low-lying areas, such as agriculture, fisheries, industry, tourism, and other facilities, especially in Alexandria and the Nile Delta. Risk areas in the Alexandria region are Mandara and El Tarh (southeastern side of the city), while in the Nile Delta region are the coastal lagoons barrier, east and west of the Rosetta City, Gamil, and the Tineh plain.

Proposed adaptive options

Various types of adaptation can be distinguished, including anticipatory and reactive adaptation, private and public adaptation, and autonomous and planned adaptation (IPCC 2001). The ability to adapt to climate change depends on several factors, including the available infrastructure, resources, technology, information, and the extent of fair distribution of resources. Recommended adaptive options to climate change and SLR along the Mediterranean coast of Egypt are Implementing Egypt's National Strategy for Adaptation to Climate Change and Disaster Risk Reduction (IDSC/UNDP 2011), setting up regulations and guidelines for coastal development incorporating the adaptation of the impacts of SLR. Engineering structures should be designed to protect high-risk areas vulnerable to inundation. Rehabilitation and strengthening of the existing engineering coastal structures, particularly on low-lying lands and areas at risk, to adapt to the likely impacts of SLR. Integrating the adaptation plans and strategies to climate change and SLR in national development plans.

Conclusions

It is therefore imperative to adopt and develop a quantitative risk assessment approach for the impact of SLR along Egypt's Mediterranean coast. Risk associated with the impacts of the SLR may be reduced and minimized provided the consideration of adequate adaptation measures. The Mediterranean coastal zone of Egypt is asymmetrically vulnerable (social and biophysical) to the projected SLR. Beaches and coastal lagoons are highly vulnerable to the potential impacts of SLR and wave overtopping during storm events.

Vulnerability of Tourism to Climate Change in the Mediterranean Coastal Area of El Hammam–ELAlamein, Egypt

https://doi.org/10.1007/s10668-021-01488-9

Abstract-Summary

The northwestern Mediterranean coast of Egypt, including the study area from El Hammam to ELAlamein, is a hub for economic and coastal tourism development. This study explores the potential impacts of climate change (CC) and sea level rise (SLR) on coastal tourism in the El Hammam—ELAlamein area. The adopted methodological framework for this study comprises the assessment of the vulnerability of the area and coastal tourism to climate change and sea level rise via the development of a digital elevation model (DEM) and inundation models, in addition to the assessment of the temperature change employing the tourism climate index (TCI). The DEM and inundation models demonstrate the vulnerability of the eastern sector of the study area to the projected SLR following the Intergovernmental Panel on Climate Change Representative Concentration Pathway (RCP 2.6 & 8.5) scenarios. Actions for adaptation and protection measures to minimize the projected adverse impact of CC are proposed to ensure the coastal development and tourism sustainability of the area. The northwestern Mediterranean coast of Egypt, comprising the area between El Hammam and El Alamein (study area), was recognized as vulnerable to the impact of CC and SLR (Frihy and El-Sayed 2013).

Introduction

Foreseen environmental issues are likely to arise in association with the CC and SLR influences on coastal and deltaic low-lying areas along the Mediterranean coast of

Egypt. Out of its belief in the consequence of climate change and its impact on the different development sectors, including the coastal areas, Egypt emphasized this issue in addition to its national environmental concern. Several studies were carried out and published on CC and its impact, including the SLR, on different sectors and areas in Egypt, particularly on its Mediterranean coastal zone (stretched for approximately 1500 km). Although the El Hammam–ELAlamein area was incorporated in the scoping study for the development of ICZM along the northern coast of Egypt (IH-Cantabria/Environics 2017), the study of the impacts of CC and SLR on this particular area was recently considered by El-Masry (2017). The present study is indicative of the extent of the vulnerability of coastal tourism in the El Hammam–ELAlamein area to the impact of CC.

Study area

The Marina El Alamein tourism agglomeration is currently the largest, modern, and sophisticated establishment in the study area, occupying approximately 26.6% of its total surface area. The importance of the Marina ELAlamein agglomeration has significantly increased the share of El Alamein town in the total number of nights and visitors not only in the study area but also in Matrouh Governorate. The average hotel occupancy throughout the year in the study area is relatively low, approximately 17% in ELAlamein and approximately 4.6% in El Hammam (IDSC-Matrouh Governorate 2011). The study area includes the El Omayed Biosphere Reserve, which extends approximately 30 km from west El Hammam to El Alamein; approximately two-thirds of the study area. In addition, the area is bordered to the west by El Alamein New Mega City, which symbolizes a core regional development area as stated in the National Strategic Plan for the northwestern coast of Egypt (2050).

Methodology

Assessment of the present and projected climatic conditions in the study area is based on the evaluation of the impact of (T) and (P) change employing the "Tourism Climatic Index (TCI)". The TCI uses seven climatic variables relevant to tourism, namely, maximum air temperature, mean air temperature, minimum relative humidity, mean relative humidity, amount of precipitation, sunshine hours, and average wind speed. The TCI is calculated as follows (Mieczkowski, 1985): (CID) is the daytime thermal comfort, (CIA) is the average thermal comfort, (R) is the precipitation, (S) is the sunshine hours and (W) is the wind speed.). Ranking of the area vulnerability to (CC and SLR) from "high" to "low" was based on indicators developed by El-Masry (20) as follows: climate variable change scenarios according to (T) & (P) and the (TCI), coastal morphology, and (SLR) (RCP 2.6) and (RCP 8.5) scenarios as indicators of the system response/impact, e.g., exposure, sensitivity, and adaptive capacity.

Results and discussion

According to Lowe and Gregory (2005), storm surge events can increase the severity of the impacts of SLR and hence increase the severity of coastal flooding in coastal areas. Frihy et al. (2010a, 2010b) indicated that there was a marked seasonality in the wave climate, with deep-water significant wave heights during the summer months averaging 0.35–0.62 m, increasing to 1.13 m during the winter months, with individual storms yielding significant wave heights of 7 m. The study area is ranked, at present, according to the results and analysis of the TCI, as very good to idealistically climatically suitable for tourism. Bigio (2009) mentioned that the SLR impact will affect Egypt, Tunisia, and Morocco through coastal erosion loss of beaches, key urban infrastructure will be at risk as well, and port structures and coastal roads will be inundated by storm surges.

Conclusions

Tourism activities in this area will be influenced by the CC and SLR impacts in terms of the change in the spatial and temporal distribution of temperatures, the partial inundation of the coastal strip, the availability of beaches for recreation, and the quality of the coastal environment. The eastern sector of the study area will most likely be inundated, and the beach will be eroded due to the projected SLR unless appropriate measures are taken and adequate resources are made available to empower coastal protection. The developed DEM and inundation models prove their effectiveness in exploring and assessing the implications and vulnerability of coastal areas and tourism to CC and SLR. Assessment of the existing adaptive measures in the area indicates their inadequacy to face the future (CC) and (SLR) and other natural risks, such as storms and tsunamis, as the existing coastal protection hard engineering constructions were established in a fragmented manner.

Change Detection in the Coastal Zone East of the Nile Delta Using Remote Sensing

https://doi.org/10.1007/s12665-010-0564-9

Abstract-Summary

The coastal zone of the Nile Delta is a promising area for energy resources and industrial activities. This coastal area witnessed several changes during the last century. A set of four satellite images from the multispectral scanner (MSS), thematic mapper (TM), and Systeme Pour l'Observation de la Terre (SPOT) sensors were utilized to estimate the spatiotemporal changes that occurred in the coastal zone between the Damietta Nile branch and Port Said between 1973 and 2007. Image processing applied in this study included geometric rectification; atmospheric correction; on-screen shoreline digitization of the 1973 (MSS) and 2007 (SPOT) images for tracking the shoreline position between the Damietta promontory and Port Said; and a water index approach for quantifying Manzala lagoon surface area change using 1973 (MSS), 1984 (TM) and 2003 (TM) images. The results showed that coastal erosion was severe near the Damietta promontory and decreased eastward; however, accretion was observed near Port Said.

Introduction

The basic principle behind using digital data is that any subtle change in land-cover/use results in a change in the radiance of that object and can be detected by satellite sensors (Mass 1999) at different spatial, spectral, and radiometric resolutions. The conversion of land use from rural to urban land causes a change in the visible portion of the spectrum (brightness) arriving at the sensor from this new land use; the changes from vegetation to non-vegetation land use cause a difference in the near-infrared (NIR) radiation (greenness) detected by the sensor; and the change in the shortwave-infrared (SWIR) reflects the change in moisture content (wetness) of an object (Lunetta and Elvidge 1998). The most common methods applied for change detection include band ratio, band differencing, principal component analysis, vegetation index differencing, and post-classification change detection. The water index concept is a new approach applied for the change detection of water bodies.

The study area

During the second half of the previous century, the coastal area of the Nile Delta experienced significant land-use/cover change due mostly to the control of the River Nile flooding regime and extensive population growth. Syvitski (2008) mentioned that between 1800 and 1900, the River Nile mouth at Damietta promontory advanced seaward by 3.0 km, while today, the entire Nile Delta coastline is retreating landward due to coastal erosion. El Banna and Frihy (2009) reported that both natural and anthropogenic factors have influenced the Nile Delta coastal area. Factors are change in the Nile sediment supply, coastal processes, land subsidence, and deterioration of natural habitats. The beaches of the Nile Delta coast are backed by coastal dunes and coastal flats with low-relief surfaces interrupted by sabkhas and salt marshes. The objectives of the present study are to apply remote sensing in mapping and addressing spatial changes that occurred along the coastal area of the Nile Delta between 1973 and 2007. Specific objectives include mapping coastline position changes between Damietta and Port Said and quantifying Manzala lagoon surface area changes between 1973 and 2003.

Materials and methods

The Landsat images (MSS and TM) were used to estimate the lagoon area change because they cover a substantial area of the Nile Delta, including the entire Manzala lagoon area. Two different image processing approaches were applied to identify the changes: (1) shoreline digitization for mapping erosion/accretion patterns along the entire Nile Delta coast between Ras El Bar (at the end of the Damietta Nile branch) and Port Said (at the entrance of the Suez Canal) using the very early (the MSS of 1973) and very last (the SPOT of 2007) images; and (2) Lake Manzala surface area changes using water index algorithms applied to the Landsat images of 1973, 1984, and 2003.

Results

Along this coastal strip, four locations of erosion (sites A, B, C, and E) and one location of accretion (site D) were identified from satellite data. The last location of erosion is encountered just after El-Fardos village and extends to 2 km west of the city of Port Said (Site E). At this 5.1-km-long strip, coastal erosion caused the shoreline to withdraw up to 0.21 km landward from its original location in 1973. The gross length of erosional strips totals 29 km, which represents 50% of the total length of the coastal area between Damietta promontory and Port Said, whereas an accretional strip occurs at 13% of this coast (7.4 km). To check the accuracy of the generated NDWI and MNDWI images, two test sites that represent shallow and/or turbid water were chosen in the raw TM images.

Discussions and conclusions

The dam has altered the Nile flow, sediment, and wastewater discharge and consequently induced alteration of the northern Nile Delta wetlands, such as the Manzala lagoon (Randazzo et al. 1998). The results of the present study agree with a previous qualitative study by El Raey et al. (1999), who used band rationing, band difference, and principal component analysis of MSS images taken in 1978, 1984, and 1990 to identify coastal changes between Damietta and Port Said and concluded that erosion is maximal near the River Nile mouth at Damietta and becomes slightly eastward. The present study shows that Manzala Lagoon lost 12,000 acres (4.4%) between 1973 and 1984, with an annual loss rate of 1,090 acres/year. Frihy et al. (1998a) estimated the lagoon area through unsupervised classification of the MSS image acquired in 1983 as 809 km^2 (compared with 1,052 km^2 in 1984 in the present study). Land subsidence threatens significant parts of the low-lying northern delta, including the Manzala lagoon (Stanley and Warne 1993).

Geo-Archeoheritage Sites Are at Risk, the Manzala Lagoon, NE Nile Delta Coast, Egypt
https://doi.org/10.1007/s12371-018-0297-9

Abstract-Summary

The Manzala Lagoon is one of the most important coastal geo-archeoheritage sites in Egypt. The lagoon originates tectonically and includes hundreds of islands, which are identified either as coastal sand ridges associated with paleo-shorelines or as riverbanks. The archeological sites at Lagan and Tinnis islands refer to ancient cities belonging to the Medieval ages. These land fillings were at the expense of the lagoon's water body, which decreased in area from 1540.85 to 853 km^2 at present; only 55.4% of the 1973 lagoon's area does exist, with an average rate of 16 km^2/year. Such processes negatively affect the lagoon ecosystem and might threaten archeological sites. The present study does not recommend the landfilling of 12.2 km^2 at the Ashtum El-Gamiel Protectorate, SW of Port Said, to construct a social housing project or roads crossing the lagoon. We recommend converting Manzala Lagoon into a tourism destination similar to Italian Venice, with the construction of an outdoor museum supported by facilities making it attractive to tourists who enjoy culture and nature. Manzala Lagoon can therefore be recognized as a geoheritage site with great geodiversity that needs to be individually studied and subjected to a geoconservation plan. The Manzala Lagoon warrants recognition as a geoheritage site characterized by great geodiversity.

Introduction

The Nile Delta in northern Egypt was formed, where the Nile River spreads out and drains into the Mediterranean Sea. The Nile River branches at the delta into two main distributaries, the Damietta to the east and the Rosetta to the west, which flow into the Mediterranean in the coastal cities of Ras El Bar and Rasheed, respectively. Three lagoons representing shallow basins (depth < 3 m, average depth of ~1 m) with brackish water are located along the northern Nile Delta coast: Edku, Burullus, and Manzala. The latter two are separated from the Mediterranean by low-lying, narrow, coastal sand barriers or bay mouth bars (El-Asmar 2002; El-Asmar et al. 2014) and are connected to the sea by protected inlets that allow water exchange.

Methods and Techniques

All images were processed using ENVI version 5.1 software. The atmospheric correction was calculated by subtracting the minimum pixel value from the image pixel matrix (dark object subtraction; Chavez (1996) to represent the clear deep water of the Manzala Lagoon. The classification process utilized all spectral bands within the ETM images, except for the thermal bands.

Results

According to the accounts of Al-Massoudi (AD 896–956) and Al-Maqrizi (AD 1364–1442), the area presently occupied by Manzala Lagoon had the most fertile soil in Egypt. According to the accounts of El-Maqrizi (1364–1442), at the beginning of the fourth century AD and before the lagoon originated, the town of Lagan was located on the overbank of the Mendisian branch of the Nile, and the city of Tinnis was built on the right bank of the Tanitic branch at a coastal sand ridge 6 km south of present-day Port Said. Remote sensing images were used in this study to detect different types of encroachments resulting from human activities along the lagoon proper and its islands and to locate the accreted sand ridges and other overbank cheniers that might represent old cities or old shorelines.

Discussion and Conclusions

Are the fish farms, which can be confused and calculated among the area of the main lagoon water body? The MNDWI index calculates the area of water irrespective of its type, whether it is the main lagoon or fish farms. The areas of main lagoon water and fish farms for Tinnis and Lagan are 27.9 and 28 km^2, respectively. The encroachments have not only diminished the water body of the lagoon through landfilling and fish farms but have also extended to most of the archaeological sites because of the search for coins and statues of antiquity that could be sold or smuggled out of the country. As the new "30 June Road," which passes 200 m off Tinnis Island, has become a reality, it is necessary to consider the best way of geoconservation of the geodiversity along the Manzala Lagoon and to safeguard its remaining archaeological sites as Tinnis from being stolen in a framework of sustainable development. The present study was proposed to shed light on Manzala Lagoon as an important geo-archeoheritage site in Egypt with characteristic geodiversity. Among such sites in the lagoon are Lagan and Tinnis, which originated as either overbank cheniers or coastal sand ridges and represent paleo-shorelines of the Late Holocene from 5000 BP to < 1000 BP. Due to the tectonic subsidence, and strong earthquake in AD 365, and associated tsunami, the seawater invaded the coastal sand ridges and the former green low-lying lands now occupied by the Manzala Lagoon. Based on the present study, we do not recommend the landfilling of ~12.2 km^2 of the lagoon that is planned by the Port Said Governorate to construct a social housing project at the expense of parts of the Ashtum El-Gamiel Protectorate. This project crosses Manzala Lagoon, 200 m off Tinnis Island.

Implications of Climate Change on the Groundwater Flow Regime and Geochemistry of the Nile Delta, Egypt
https://doi.org/10.1007/s11852-015-0409-5

Abstract-Summary

Environmental isotope analyses in conjunction with hydrogeochemical investigations and a tentative review of paleoclimatic sea level changes are carried out to fingerprint the implications of climatic changes on the groundwater flow regime and geochemistry in the Nile Delta. Following up on the footprints of the groundwater flow history, it is observed that the main groundwater aquifer of the Nile Delta in the Pleistocene was drained and refilled with Nile water several times due to the eustatic sea level propagations between dry and wet periods. The present-day groundwater flow regime could be affected by the latest Holocene phase of climate change, during which no significant dramatic sea level changes were recorded. In accordance with the latest active sea level rise stage in conjunction with delta subsidence, a contagious groundwater level rise with a recent order of 3 cm/year is taking place, leading to the formation of several lake-like lagoons, water logging, and soil salinization along the coastal plain and the eastern lowlands. The Nile Delta is expected to suffer extreme soil salinization and gradual merging under groundwater logging and seawater transgression, especially along the eastern coastal zone, which suffers a high subsidence rate of approximately 5 mm/year. In previous studies, our findings show that the present groundwater composition and salinity in the Nile Delta aquifers cannot be attributed to a recent seawater intrusion.

Introduction

Global climate change has already had observable effects on the environment. Effects that scientists had predicted in the past would result from global climate change are now occurring: loss of sea ice, accelerated sea level rise, and longer, more intense heat waves (IPCC 2007). The northern coastal zone of the Nile Delta is generally low land and is consequently vulnerable to direct and indirect impacts of sea level rise due to climate change, particularly inundation (Hassaan 2013). The global glacial-eustatic variation in sea level was the most effective paleoenvironmental control on the Nile Delta Formation, especially during the Quaternary (Zaghloul et al. 1979; Fairbanks 1989; Geirnaert and Laeven 1992; Diab et al. 1997; Geriesh 2000; Hassaan 2013; El-Asmar et al. 2015). The main objectives of the present study are to trace the different groundwater types, map their geographic distributions, and decipher their physico-chemical processes, origin, and recharge history to identify the potential implications of climate changes on the groundwater flow regime and geochemistry in the Nile Delta region.

History of paleoclimatic—sea level changes

This delta is characterized by its fine-grained sediments and constitutes the impermeable base of the Quaternary aquifer; C) the Prenile phase, at which the delta acquired its maximum enormous size (approximately four times the present-day delta size), occurred during the Middle Pleistocene. The following climatic-sea level cycles could be inferred; this cycle marks the surface of the Nile Delta at the time of seasonal beta, gamma, and Neonile (approximately 21,500 y BP). These valleys continued the erosion stage of the previous alpha Neonile and eroded most of the upper Prenile loose sediments (approximately 45 m above present sea level) (Said 1993; El Fawal and Shendi 1991) and left only the intervalley parts as sand islands, some of which disappeared later under the late Holocene Nile sediments while others attained higher altitude (reaches up to 45 masl) and are still observed in different parts of the present delta.

Implications of climatic changes on the groundwater flow regime

In spite of the abovementioned discussions and the obtained results, the groundwater aquifers in the Nile Delta region have been discharged and refilled by Nile water of different sources along its long history starting from the complete performance of its shape at the Middle Pleistocene until the present day due to severe climatic and sea level changes. During this humid period, considerable discharge from the northern part of the aquifers occurred, and most of the deep Nile distributaries at that time acted as effluent channels along the northern parts of the delta. 2- Refilling of the Delta aquifers could then have taken place during the next pluvial phase of the early Holocene period (approximately 8500–5000 years BP), at which time the sea level was raised to approximately 25 m below the present level. Continued sea level rise at the end of this phase helped recharge the northern parts of the Nile Delta aquifers.

Discussion and Conclusions

Following up on the footprints of the groundwater flow history, it is observed that the main groundwater aquifer of the Nile Delta in the Pleistocene was drained and refilled with Nile water several times due to the eustatic sea level propagations between dry and wet periods throughout this history. We hypothesize that this pluvial period was the principle phase of refilling the Pleistocene aquifers of the Nile Delta and its surrounding desert fringes. (B) Aquifer depletion and deterioration-arid phase prevailed during the invasion of the Holocene Sea to the low-laying lands and the mouths of the ancient Nile branches to a considerable distance inland and formed a considerable sequence of salt lakes and playa deposits during the time slice from 5000 to 3500 y BP. Naturally, imported Nile water is thought to be the main source of recharge for the whole groundwater budget in the Nile Delta and surrounding desert fringes during the last 7500 years, which reflects progressively prevailing arid conditions in the region during this long history.

Vulnerability Assessment of African Coasts to Sea Level Rise Using GIS and Remote Sensing
https://doi.org/10.1007/s10668-020-00639-8

Abstract-Summary

This study attempts to fill the lack of studies and assessments of the African coasts by developing an SLR vulnerability assessment. The study customized the coastal vulnerability index "CVI" to include seventeen parameters grouped into vulnerability pillars: exposure, sensitivity, and resilience. The study applies the CVI method for the African coastal zone using GIS and remote sensing. The study has followed the case study approach on the continental level of Africa, and the results have classified coastal areas into different degrees of vulnerability. Application of the CEI equation showed that approximately 40% of the African coast ranges from moderate to very high exposure, the CSI equation showed that 75% of the African coast ranges from moderate to very high sensitivity, the CRI equation showed that 55% of the African coast ranges from moderate to very high resilience, and the CVI equation showed that 35% of the 26,000 km length of Africa's coasts are vulnerable to SLR.

Introduction

It is expected that SLR on African coasts will increase by 10%, on average, over global rise (IPCC, R5 88). The African coast represents a belt for 33 countries and 7 islands, representing very rich habitats and important settlements, which are affected by the hazards of sea level rise (SLR), such as storm surge, high tidal ranges, hurricanes, and cyclones. Many approaches associated with different methods have been used to assess the coastal community's vulnerability and to identify the appropriate adaptation of SLR policies and strategies. In light of that, the coastal vulnerability index (CVI) is a flexible and integrated method used widely to assess coastal vulnerability. It is an analytical approach to assess coastal vulnerability based on a group of different physical, social, and economic variables.

Methodology and study area

The used method provides an initial assessment and a relative measure of the system's vulnerability to SLR impacts at the regional scale by visualizing the coastal vulnerability of the study area's cells with a specific way to be a useful adaptation tool for spatial planning and coastal management strategies. Index-based integrated studies can be adopted in different disciplines and at multiple scales, but they should be performed on the same coastal parameter cells (parcel resolution) to rank homogeneous vulnerable cells for each parameter and to establish the real hotspot areas to finally define the priorities of intervention. The study is based on multicriteria decision analysis to rank these parameters. The model used in this study is based on uncertainty (in model structure and parameter selection) to assess the coastal vulnerability pillars (exposure, sensitivity, and resilience) by using their dimensions.

Ranking system of the selected parameters

At the same time, there is a variation in pillar groups: exposure parameters (topography, slope, proximity to coast, urban population percent, land cover, accessibility, soil type, and elements at risk indicator), sensitivity parameters (vegetation percent, vegetation type, natural resources sensitivity indicator, and growth rate), and parameters of adaptive capacity by human interference (human capital, financial capital, institutional capital, infrastructure, and household technology indicators). The ranking system based on a quantitative variation of the parameters is used in the case of topography, slope, proximity to the coast, urban population percent, accessibility, vegetation percent, growth rate, and the indicators used. The very highly vulnerable area, which is based on topography parameters, is classified as the area from the shoreline to contour line +5 because the DEM data accuracy is less than 3.7 m. The highly vulnerable area is classified as the area from +5 to +10 m because some parts of the sand dune belt exist in this zone that protect the inland areas from different SLR hazards.

Mapping of vulnerability parameters

Topography is a dominant and main parameter used in assessing vulnerability to SLR because any area below or above mean sea level will be impacted by an inundation risk because of any rise in sea level. A high population growth rate will increase the future number of people likely to be affected by the impacts of SLR; therefore, areas with lower growth rates will have less vulnerability than those with higher growth rates. The high rate of technology is an indicator of the high quality of life, and thus, it raises the capacity to cope and face the impacts, as in North and South Africa (Jahn et al. 2006).

Results and discussion

The selected parameters to assess the coastal vulnerability to SLR consequences on the African coasts reflect the spatial circumstances related to the natural and built environment there. The model was developed to determine highly vulnerable areas (hot spot areas) in the coastal zone of Africa, which allows decision-makers to manage the risk by the preadaptation process. The areas exposed to SLR that ranged from moderate to high degrees are along the coasts of Tunisia, Algeria, Morocco, Senegal, Benin, Togo, Ghana, Ivory Coast, Liberia, Sierra Leone, Guinea Bissau, Nigeria, Kenya, southern Somalia, and northwestern Libya. Moderate vulnerable areas exist in some coastal parts in Kenya, Tanzania, Mozambique, Angola, Zaire, Cameroon, Nigeria, Ivory Coast, Liberia, Sierra Leone, Guinea, and Senegal.

Conclusion

Application of the CEI equation showed that approximately 40% of the African coast ranges from moderate to very high exposure, the CSI equation showed that 75% of the African coast ranges from moderate to very high sensitivity, the CRI equation showed that 55% of the African coast ranges from moderate to very high resilience, and the CVI equation showed that 35% of the 26,000 km length of Africa's coasts are vulnerable to SLR. While North Africa is highly exposed and highly sensitive to SLR impacts, it has low vulnerability due to its high resilience against SLR hazards. To decrease the vulnerability degree in coastal African communities, the exposure and sensitivity criteria in these communities to hazards, especially SLR, should be decreased, and the resilience criteria should be increased.

GIS-Based Risk Assessment for the Nile Delta Coastal Zone Under Different Sea Level Rise Scenarios Case Study: Kafr EL Sheikh Governorate, Egypt

https://doi.org/10.1007/s11852-013-0273-0

Abstract-Summary

Global average sea level is expected to rise by 2100 due to global warming between 0.18 and 0.59 cm. Such a rise in sea level will significantly affect coastal areas due to the high concentration of natural and socioeconomic activities and assets located along the coast. The northern coastal zone of the Nile Delta is generally lowland and is consequently vulnerable to direct and indirect impacts of sea level rise (SLR) due to climate change, particularly inundation. Despite the uncertainty associated with developed scenarios for climate change and expected SLR, there is a need, according to a precautionary approach, to assess and analyze the impacts of SLR. Such an analysis can contribute significantly to the development of an integrated approach to address the impacts of SLR. The objective of this paper is to assess and spatially analyze the risks of expected sea level rise (SLR), in particular inundation, and its implications up to 2100 in Kafr El Sheikh Governorate, Egypt, using GIS techniques. A GIS was developed for the study area and then utilized to identify the spatial extent of those areas that would be vulnerable to inundation by SLR. No significant difference was noticed between the two scenarios in terms of the spatial extent of SLR impacts. It was found that approximately 90.13% of vulnerable areas are actually less exposed to the risks of SLR due to the existence of a number of man-made features that are not intended as protection measures, e.g., the International Coastal Highway, which can be used to limit the areas vulnerable to inundations by SLR. There is a need to profile socioeconomic and demographic conditions in the coastal zone of the study area up to 2100 to estimate the potential impacts of SLR on future socioeconomic settings by 2100.

Introduction

Such an assessment is based on both the B1 and A1FI scenarios of global SLR up to 2100 and the rates of land subsidence in the study area. Based on these SLR scenarios, the study estimated that approximately 21.6–26% of the Nile Delta habitat would be lost due to inundation, which would lead to the displacement of approximately 19–24% of the Delta population. El Raey et al. (1995) studied the impacts of SLR on the city of Alexandria under two hypothetical scenarios of 0.5 and 2.0 m and estimated that approximately 57% and 76% of its total area would be vulnerable to inundation, respectively. It can be argued that despite their diverse estimates, previous studies agreed that the coastal zone of the Nile Delta would be highly vulnerable to inundation due to expected SLR as a result of climate change.

The study area

Kafr El Sheikh Governorate, the study area, is located in the northern part of the Nile Delta and extends from 30° 59′ 38″ to 31° 36′ 00″ latitude and from 30° 21′ 40″ to 31° 18′ 40″ longitude. The governorate is bordered by the Mediterranean Sea to the north, with a coastline of approximately 100 km, Rosetta Nile Branch and Behaira Governorate to the west, Dakahleya Governorate to the east, and Gharbeya Governorate to the south. One of the main geographical features of the coastal area of the governorate is El Borolos Lake, which is located in the northern parts of the Nile Delta. Kafr El Sheikh Governorate, as a part of the Nile Delta, is highly vulnerable to SLR impacts, which may lead to increased coastal erosion, overtopping of coastal defenses and increased flooding, damage to urban centers and infrastructure, retreat of barrier dunes, increased soil and lagoon water salinity, and decreased fisheries production (MSEA 2001; UNDP-RBAS 2012).

Materials and methods

This means that estimating the relative SLR for each part of the study area entails developing synoptic coverage for the study area representing total land subsidence up to 2100. The relative sea level rise by 2100 in each part of the study area was estimated by adding the total subsidence raster surface produced in the previous step to the average global SLR according to the B1 and A1FI scenarios, which was 28 and 43 cm, respectively. Using select by location tools, polygons representing low-laying land, previously identified, that are spatially adjacent to the sea and in direct contact with coastline (touch the boundary of sea) were selected and exported to create a new polygon feature class representing areas that were susceptible to inundation due to SLR.

Results and discussion

Of the expected relative SLR, it was found that approximately 844.08 and 915.29 km^2 of the total area of the Kafr El Sheikh Governorate would be vulnerable to inundation due to SLR under the B1 and A1FI scenarios, respectively. These five districts can be classified into three main categories according to the area vulnerable to inundation due to SLR by 2100 as follows: Highly vulnerable districts, including Metobas and SediSalim, are located on the eastern side of the governorate, with vulnerable areas exceeding 50% of their total areas. This means that approximately 11.6% and 14.0% of the total roads of the Kafr El Sheikh Governorate would be lost to inundation according to the B1 and A1FI scenarios, respectively. This, in turn, may suggest that valuable areas vulnerable to inundation by SLR in the Kafr El Sheikh Governorate could be smaller in reality than estimated.

Conclusion

The assessment conducted in this study, which focused on the risks of inundation due to SLR on present assets in Kafr El Sheikh Governorate, showed that more than 22% of the total governorate area would be vulnerable to inundation according to the B1 and A1FI Scenarios. There are already a number of man-made features that were not intended as coastal protection works, for instance, the International Coastal Highway, that can be used to decrease the areas vulnerable to inundations due to SLR. This, in turn, means that areas vulnerable to inundation by SLR in the governorate could yet be smaller than estimated in this paper. If no careful consideration of the areas vulnerable to inundation due to SLR is taken into policy and/or decision-making, potential impacts and damages could increase significantly.

Mapping Coastal Erosion at the Nile Delta Western Promontory Using Landsat Imagery

https://doi.org/10.1007/s12665-011-0928-9

Abstract-Summary

A set of six Landsat satellite images 5–9 years apart was used in a post-classification analysis to map changes that occurred at Rosetta promontory between 1973 and 2008 due to coastal erosion. To estimate the quantity of land loss in terms of coastal erosion, a supervised classification scheme was applied to each image to highlight only two classes: seawater and land. The results showed that Rosetta promontory lost 12.29 km^2 of land between 1973 and 2008, and the shoreline withdrew southward by approximately 3.5 km due to coastal erosion. Most land loss and shoreline retreat occurred between 1973 and 1978 (0.55 km^2/year and 132 m/year, respectively). These structures have considerably contributed to reducing coastal erosion; however, they have promoted downdrift erosion.

Introduction

Change detection of satellite data ranges from visual comparison to complex image processing schemes. Post-classification change detection is a famous technique applied to address thematic changes in satellite data. The present Nile Delta has two promontories: the eastern is at Damietta, and the western is at Rosetta. The majority of research carried out upon monitoring coastal erosion at the delta promontories, particularly Rosetta, was extracted from topographic maps, aerial photographs, and beach profile studies. Blodget et al. (1991) used MSS data to monitor the shoreline in the Rosetta region between 1972 and 1987. Frihy et al. (1994) conducted a shoreline change study using MSS images between 1972 and 1991 and observed that the headland retreat rate approached 71 m/year. The objective of the present study is to update and address the change that occurred at the Rosetta headland until 2008, taking into account the impact of the constructed coastal protection measures on erosion/accretion patterns.

Materials and methods

To make an accurate change detection using the MSS, TM, and ETM+ images, the two MSS images, which have a spatial resolution of 60 m, were resampled to 30 m pixel size, so that all the satellite images have the same spatial resolution (30 m). Each classified image was then recoded into land and water. A total of 50 random pixels were selected at each land-cover map using the stratified random approach and compared with the original pixels in the corresponding images. One important advantage of post-classification change detection is that the images are classified separately, which minimizes the problem of radiometric normalization between dates and lowers the amount of preprocessing work (Song et al. 2001; Wang et al. 2009). The change in the total land area from each classified image was estimated, and the change matrix was obtained.

Results and discussion

The maximum land loss rate (erosion) occurred during 1973–1978 (0.55 km^2/year), whereas the minimum occurred during 1999–2008 (0.03 km^2/year) due to the construction of an integrated coastal protection system. The construction of the delta barrages resulted in coastal erosion of Rosetta by 20 m/year during the period 1941–1964 (El Sayed and Mahmoud 2007). The rate of erosion at the headland was significantly lowered due to the construction of two seawalls along the eastern and western margins of the headland during 1986–1987; however, coastal erosion was observed afterward at the unprotected beaches. Mikhailova (2001) mentioned that the main factors contributing to the washout of the coastline of the Nile Delta are the human-induced cutoff of sediment discharge through the River Nile, sea waves, eustatic sea level rise, and land subsidence. This area is periodically inundated during winter; consequently, a small rise in sea level will accelerate coastal erosion and seriously affect any land use in the region.

Conclusions

Mapping coastal change using satellite remote sensing asserts the advantage of using this tool over tedious routine ground surveys. Field truthing is an indispensable task to verify remotely sensed data. The main conclusion of this study is that the headland of the Rosetta promontory witnessed severe coastal erosion between 1973 and 2008. Although coastal protection works have helped protect the coast, they have some erosion consequences along their margins and at their downdrift sides.

DEMs and Reliable Sea Level Rise Risk Monitoring in the Nile Delta, Egypt

https://doi.org/10.1007/s43621-020-00006-7

Abstract-Summary

This research aims to evaluate the accuracy and reliability of utilizing global digital elevation models (DEMs) in the risk monitoring of sea level rise (SLR) in the Nile Delta region, Egypt. Based on novel proposed evaluation indicators, it has been found that the reliability of three investigated global models (namely, SRTM1, ASTER, and EarthEnvi-DEM90) in delineating sea level rise flooded areas is less than 20%. It is concluded that the utilization of global DEMs may not be optimal in the planning and management of coastal areas in the Nile Delta region. The expected findings would be crucial for decision-makers not only in Egypt but also in other regions of low-lying deltas.

Introduction

The rise in sea level could influence coastal regions in several hazardous ways, including increasing the frequency and severity of coastal flooding, increasing the length and energy of waves, increasing the salinity of groundwater, and raising the groundwater surface elevation (Raymond et al. 2018). Several hazardous impacts of sea level rise have been investigated worldwide, particularly in lowland deltas, which necessitate the development of risk assessment adaptation planning (Abutaleb et al. 2018). In the context of seal level rise hazard monitoring, several studies and even practical projects depend on the utilization of global DEMs in Egypt and other countries. Rather than the pointwise evaluation of GDEM vertical accuracy, this paper provides a novel approach of using several models with variable spatial resolutions and vertical accuracies for the accurate monitoring and assessment of sea level rise impacts in Nile Delta coastal areas.

Study area and materials

Dawod et al. (2019) investigated relative and absolute sea level rise in the Nile Delta area based on tide-gauge records and GNSS-based datasets. Three global DEM models (GDEMs) were utilized in the current study, namely, the Shuttle Radar Topography Mission (SRTM), EarthEnv-DEM90, and Advanced Spaceborne Thermal Emission and Reflection Radiometer (ASTER). SRTM is a global DEM that has two versions: SRTM1 with a spatial resolution of 1 arc-second (30 m) and SRTM3 with a 3 arc-second resolution (downloaded from https://earthexplorer.usgs.gov/). They represent different spatial resolutions (1 and 3 arc-seconds), which would help assess the sensitivity and precision of GDEMs to spatial variations in the process of SLR impact monitoring (Kotb et al. 2017). The SRTM1 and ASTER have been widely utilized in risk assessments due to sea level rise.

Methodology and processing

This high-precision level enables accurate estimation of SLR-based inundated regions and, subsequently, truthful estimation of GDEM corresponding results and reliability. This methodology could be applied in other low-lying deltaic areas worldwide to investigate GDEM reliability for SLR monitoring and assessment in coastal regions. These two results give an overall view of the performance of the utilized global models. Such results only present quantitative measures regarding the precision of several DEM models with varying levels of accuracy and spatial resolution. A spatial perspective utilizing ArcGIS 3D analysis and spatial analysis tools is still desired for the accurate mapping of inundated regions due to SLR. Since the local DEM is accurate, the overlapping parts of the hazardous areas between the local model and a particular global model could be considered as a truth index for that model.

Results and discussion

Compared to the local SRI18-DEM, the EarthEnv-90DEM model has exaggerated the anticipated inundated regions by approximately 111%. This comparison demonstrates the reliability of each utilized GDEM in delineating the spatial distribution of its corresponding flooded regions compared to the truthful SRI18-DEM results. It can be concluded that the total areas of inundated regions cannot be considered an optimum measure of global DEM performance regarding SLR impacts. It can be noticed, from this table, that the three GDEMs identify very small portions of the actual inundated regions resulting from the accurate local DEM. It can be concluded that the findings herein are superior to those of previous studies in the Nile Delta region due to the utilization of the local high-accuracy DEM for the first time in Egypt.

Conclusions

The current research investigates the accuracy and reliability of some global DEMs over the Nile Delta coastal region, namely, EarthEnvi-90, SRTM1, and ASTER. Such an evaluation has been carried out based on a local accurate DEM of the area generated from terrestrial GPS/leveling surveying of approximately 300 square kilometers. According to the proposed indices for investigating the trustworthiness of DEMs, it has been concluded that the reliability of the three global models in delineating sea level rise flooded areas is less than 20%. Such findings play a major role for decision-makers to avoid the utilization of global DEM models in the accurate planning and integrated management of coastal areas in the Nile Delta region. Based on available datasets, the results stress that a local accurate and up-to-date terrestrial DEM is necessary for precise risk assessment of sea level rise in Egypt and other developing countries.

Vulnerability of the Nile Delta Coastal Areas to Inundation by Sea Level Rise
https://doi.org/10.1007/s10661-012-3050-x

Abstract-Summary

Such a rise in sea level will significantly affect the coastal area of the Nile Delta, which consists generally of lowlands, densely populated areas, and accommodates a significant proportion of Egypt's economic activities and built-up areas. The Nile

Delta has been examined in several previous studies, which worked under various hypothetical sea level rise (SLR) scenarios and provided different estimates of areas susceptible to inundation due to SLR. The paper intends, in this respect, to identify areas, as well as land use/land cover, susceptible to inundation by SLR based upon most recent scenarios of SLR, by the year 2100 using GIS. The results indicate that approximately 22.49, 42.18, and 49.22% of the total area of the coastal governorates of the Nile Delta would be susceptible to inundation under different scenarios of SLR. It was found that 15.56% of the total area of the Nile Delta would be vulnerable to inundation due to land subsidence only, even in the absence of any rise in sea level. This suggests that the inundation impact of SLR on the Nile Delta is less than previously reported. The Nile Delta region, which occupies an area of 23,850.76 km^2 (El Nahry and Doluschitz 2010), contains the most fertile land in Egypt. The Nile Delta is very heavily populated, with approximately 1,600 inhabitants/km^2, and contributes 30–40 and 60% of agriculture and fish catch production, respectively (Frihy 2003). The Nile Delta shoreline extends from Alexandria in the west to Port Said in the east, with a total length of approximately 240 km. It was found that various types of roads were susceptible to inundation due to SLR, ranging between 4,283.22 and 11,805.74 km. It was found that a total length of 3,753.89 and 10,606.36 km of canals and drains are expected to be inundated by SLR.

Introduction

Another study covering the whole Nile Delta coastal area, assuming a mean SLR ranging between 1.6 and 2.3 mm/year, suggested that approximately 30% of the Nile Delta coast would be vulnerable to SLR (Frihy 2003). El Nahry and Doluschitz (2010) studied the impacts of hypothetical SLRs of 1.0, 1.5, and 2.0 m and suggested that 28.93, 35.33, and 50.78% of the coastal area of the Nile Delta would be lost, respectively. It can be argued that the quite diverse estimates of areas vulnerable to inundation in the Nile Delta coastal area could be due to the wide range of hypothetical scenarios for SLR employed, which ranged between 0.28 and 3.32 m. Furthermore, these previous studies did not specify the inland spatial extent of their studies, which means that the percentage result figures are not comparable.

Expected sea level change

Chust et al. (2010) attempted to estimate future changes in the level of the Mediterranean Sea, a study on sea level changes in the Bay of Biscay during the twenty-first century suggested that sea level is projected to rise in the range of 28.5 and 48.7 cm up to 2099 under A1B and A2 IPCC scenarios. Since different parts of the Nile Delta have been experiencing varied land subsidence rates, ranging between 0.5 and 4.5 mm/year, they would be susceptible to different levels of relative SLR. Projected relative SLR to be experienced in the northern section of the Nile Delta would range, up to 2100, between 64–104, 145–185, and 205–245 cm for the highest projected SLR estimates according to the A1FI, Rahmstorf, and Pfeffer, respectively.

Results and discussion

The total areas of the Nile Delta that would be vulnerable to inundation, according to land subsidence only, in the A1FI, Pfeffer, and Rahmstorf highest projected SLR were found to be approximately 2,672.0, 4,006.24, 7,335.97, and 8,788.78 km^2, respectively. The six coastal governorates of the Nile Delta can be classified into two main categories according to the area vulnerable to inundation due to SLR by 2100 as follows: Highly vulnerable governorates, including Port Said Damietta and Kafr El Sheikh, located on the eastern side of the Nile Delta. It was noted, from the spatial extent of SLR impacts, that governorates located on the eastern side of the Nile Delta were more vulnerable to inundation by SLR. This may suggest that areas vulnerable to inundation by SLR in the Nile Delta area could be smaller than estimated.

Conclusion

Assessment of the physical impacts of inundation due to SLR on present assets in the Nile Delta coastal zone showed that a large proportion of the Nile Delta would be vulnerable to inundation. The impacts due to land subsidence were found to be significant, where approximately 15.56% of the total areas of the Nile Delta coastal governorates would be susceptible to inundations even in the case of no SLR. Future development plans should make decisions on the spatial extent and scale of development in coastal areas to limit the potential overlap between such development and SLR impacts.

Investigation of Potential Sea Level Rise Impact on the Nile Delta, Egypt, Using Digital Elevation Models

https://doi.org/10.1007/s10661-015-4868-9

Abstract-Summary

The future impact of sea level rise (SLR) on the Nile Delta region in Egypt is assessed by evaluating the elevations of two freely available digital elevation models (DEMs): the SRTM and ASTER-GDEM-V2. To provide a more accurate assessment of the future SLR impact on the Nile Delta's land and population, this study corrected the DEM elevations by using a linear regression model with ground elevations from a GPS survey. The DEM's vertical accuracies were assessed using GPS measurements, and the uncertainty analysis revealed that the SRTM DEM has a positive bias of 2.5 m, while ASTER-GDEM-V2 showed a positive bias of 0.8 m. The future inundated land-cover areas and the affected population were illustrated based on two SLR scenarios of 0.5 m and 1 m. The SRTM DEM data indicated that 1 m SLR will affect approximately 3900 km^2 of cropland, 1280 km^2 of vegetation, 205 km^2 of wetland, and 146 km^2 of urban areas and cause more than 6 million people to lose their houses. The overall vulnerability assessment using ASTER-GDEM-V2 indicated that the influence of SLR will be intense and confined along the coastal areas. The uncertainty analysis of the DEM elevations revealed that the ASTER-GDEM-V2 dataset product was considered the best to determine the future impact of SLR on the Nile Delta region. To evaluate the DEM elevations, several evaluation points were sampled from the two DEMs corresponding to the actual GPS locations. The binary images representing different SLR scenarios are multiplied by the reclassified MODIS LULC data and the future population count. These estimates of land area and at-risk populations are valuable and essential for future risk mitigation and preparedness for communities along the Nile Delta coastal areas.

Introduction

To provide the Nile Delta's coastal community with the most accurate estimates regarding SRL impacts, ground-controlled elevations should be incorporated with the analyzed digital elevation data. We are investigating the application of DEM elevation to study the impact of SLR on the Nile Delta region. The Nile Delta is an area of the world that lacks detailed ground truth data and monitoring stations. Despite the economic importance of the Nile Delta, it could be considered one of the most data-poor regions regarding the available information. As the SLR problem along the Nile Delta through the twenty-first century is imminent, to have a better perspective about its impacts, we should employ the most validated elevations with the highest available ground resolution. Elevation acts as the main parameter to determine the SLR impact on the Nile Delta region.

Data and methodology

The datasets that are used in this research are two DEM datasets: the 3 arc-second SRTM data with approximately 90 m ground resolution and the 1 arc-second ASTER-GDEM-V2 data with approximately 30 m ground resolution. Global Positioning System (GPS) elevation points were collected to evaluate the accuracy of the DEM datasets. The Delta's population information is derived from two main sources: the Gridded Population of the World version 3 (GPWv3) datasets and the census data from the Egyptian Central Agency for Public Mobilization and Statistics (CAPMAS 2014). It can be concluded that the errors varied among different DEM products according to the nature of the study region and the type of biophysical land cover (Guth 2010). A linear regression fitting between the DEM datasets and the GPS reference elevations is performed to eliminate the DEM errors and to adjust the digital elevation data according to the same benchmarks.

Results and discussion

The original SRTM data analysis indicated that at the 0.5 m-based SLR scenario, the potential affected population would be approximately 3 million people. The adjusted SRTM data indicated, however, that at 0.5 m SLR, the affected land cover would be 98 km^2 of vegetation, 14 km^2 wetlands, 804 km^2 of cropland, and 22 km^2 of urban area. The original ASTER-GDEM-V2 data analysis showed significantly different results when compared to the original and adjusted SRTM DEM dataset results. The two original DEM datasets showed distinctly different results in terms of the areas vulnerable to SLR effects when the same analysis steps were applied. The comparison of the vertical resolution of the SRTM data and ASTER-GDEM-V2 data was accomplished by calculating the elevation mean difference, and the DEM vertical accuracy was represented by the ME.

Conclusion

These two DEMs are used to determine the SLR impact on the Nile Delta region. The original and corrected DEMs were integrated with the MODIS land-cover types and gridded population datasets for SLR affect estimation on Nile Delta land and population. The SRTM and ASTER-GDEM-V2 DEM-based comparison showed substantially different SLR impact results on the Nile Delta land and population. Using the original SRTM DEM data, the expected SLR impact will be significant, while the original ASTER-GDEM-V2 DEM data revealed that inundation would affect a small portion of the coastal area. The difference between the resolutions of the two DEM datasets and the nature of the backscattered radar beams are the main factors for revealing quite different results about SLR impact in the Nile Delta region.

1.3.2 Section 2 (Coastal Erosion, Shoreline Retreat, Coastline Dynamics, Geomorphology)

Machine generated keywords: *headland, ras bar, beach, breakwater, alexandria, ridge, bar, plain, damietta, damietta harbor, seafloor, rosetta damietta, abu qir.*

Alexandria-Nile Delta Coast, Egypt: Update and Future Projection of Relative Sea Level Rise
https://doi.org/10.1007/s12665-009-0340-x

Abstract-Summary

Estimated rates from five tide gauges are variable in terms of magnitude and temporal trend of rising sea level. Analysis of historical records obtained from tide gauges at Alexandria, Rosetta, Burullus, Damietta, and Port Said show a continuous rise in mean sea level fluctuating between 1.8 and 4.9 mm/year; the smaller rate occurs at the Alexandria harbor, while the higher rate occurs at the Rosetta promontory. These uneven spatial and temporal trends of the estimated relative sea level rise (RSLR) are interpreted with reference to local geological factors. Holocene sediment thickness, subsidence rate, and tectonism are correlated with the estimated rates of relative sea level change. Projection of the averaged sea level rise trend reveals that not all the coastal plains of the Nile Delta and Alexandria are vulnerable to accelerated sea level rise at the same level due to wide variability in the land topography, which includes low-lying areas, high-elevation coastal ridges and sand dunes, accretionary beaches, and artificially protective structures. The interaction of all aspects (tectonic regime, topography, geomorphology, erosion rate, and RSLR rate) permitted us to define risk areas that are very vulnerable to the impacts of sea incursion due to accelerated sea level rise.

Introduction

In deltaic areas of low-lying coastal plains, climate change, particularly sea level rise, is already considered an issue of great importance, while on highly elevated land, this issue is minor. Sea level curves for the Mediterranean are commonly presented as relative sea level (RSL) curves (Pirazzoli 1991) because they incorporate absolute world (eustatic) sea level rise (~1 mm/year), local land motion and compression/ dewatering of prodelta and delta-front sediments. The more important factor for local sea level rise is generally the effect of subsidence or uplift within a coastal area. An area with tectonic uplift equal to the eustatic rise of sea level will experience no change in relative sea level (RSL), whereas an equal rate of subsidence will result in a "doubled" sea level rise (Milliman et al. 1989). We examine the vulnerability of the coastal plain of Alexandria and the Nile Delta to the coastal impacts of climate change, including accelerated sea level rise.

Study area

Local subsidence in the Nile Delta coastal plain, ranging from ~1 to 5 mm/year, was estimated from numerous radiocarbon-dated cores recovered at the coastal plain of the Nile Delta (Stanley 1990). The sector of greatest annual mean subsidence (4–5 mm/year) is the area close to the Manzala lagoon and the northeastern sector of the delta (Stanley 1988b). Unlike using tide-gauge records, sea level change along the study area has been evidenced by comparing the elevation of submerged archeological remains (submerged cities, coastal roads, harbors, and other structures) with present mean sea level (Ibrahim 1963, 117; El Sayed 1988, 118; Stanley et al. 2006). Submergence of archeological sites to depths of 5–7 m is recorded in Abu Qir Bay, and higher values (lowering to 5 mm/year) are recorded along the NE corner of the Nile Delta across the Manzala lagoon area.

Artificial and natural sea-defense system

Interactions between shoreline geomorphology, backshore topography, erosion/ deposition events, subsidence, coastal processes, and engineered structures along the study coastline were analyzed by Frihy (2003) to determine the degree to which the Nile Deltaic plain is susceptible or vulnerable to sea level rise and erosional processes. Fifteen percent of the delta coastline is artificially protected by engineering structures, 30% is exposed with no protection, and 55% is naturally protected by coastal dunes and accreted beaches along embayments and promontory saddles. Like other low-lying deltas, the coast of the Nile Delta has been designated as a vulnerable zone to beach erosion because of ongoing coastal processes and accelerated sea level rise, in addition to human influences. For the aforementioned manufactured engineering structures, a natural defense system provides a natural mechanism for protecting the coastal plain of the Nile Delta and Alexandria against beach erosion and SLR.

Factors controlling sea level changes

Stanley and Goodfriend (1997) reported thick Holocene sediments of 46.5 m and 42 m in cores recovered at Port Said and Port Fouad, respectively, which indicate similar subsidence rates of ~3.98 mm/year and annual relative sea levels of ~5 mm/year from 8,540 years ago to the present. Rates of subsidence across the northern delta range from <1 to >4 mm per year in the northeastern sector. Subsidence has been considerably lower in the westerly direction, ranging from ~1 mm/year near Baltim resort on the north-central coast to 0.5 mm/year farther to the west, in the NW Nile Delta and the Alexandria region (Stanley and Warne 1993). These structural trends indicate that the main cause of subsidence in this northern Nile Delta region is ongoing faulting, as well as downwarping, of the underlying 3,000 m of late Miocene to Quaternary sequences.

Methods

Data comprise the annually averaged tide-gauge records calculated from the hourly heights of sea level taken (with discontinuance) at Alexandria (1944 through 2006; 60 years), Abu Qir (1992 through 2005; 14 years), Rosetta (1969 through 2008; 17 years), Burullus (1976 through 2006; 24 years), Damietta (1990 through 2007; 13 years), and Port Said harbor (1926 through 2000; 47 years). The statistical relationships between the rate of sea level changes estimated in the present study versus Holocene sediment thickness and land subsidence are also correlated. An attempt has been made to project the average rate of the estimated sea level rise on the current land surface topography of Alexandria and the Nile Delta coastal plain. Relative proportions of areas of topographic features were calculated relative to the study coastal plain spanning between the coastline and up to the 4-m contour above mean sea level, within approximately 50 km of the coast.

Discussion

We attempt to link the updated rates of sea level changes determined at the coastal plain of Alexandria and the Nile Delta with the geological processes acting in the study area (thickness of Holocene sediment that is susceptible to compaction and dewatering, subsidence rates, and tectonic activity). Low-lying areas (below sea level of zero up to 1 m elevation) are dominant across much of the coastal plain of the Nile Delta and cover 27.97% of the coastal plain (~13,610.41 km^2), and they may be susceptible to serious land loss due to sea incursion unless protection control measures are implemented, such as the barrier at Manzala lagoon. Based on the estimated beach erosion caused by coastal processes and geological factors (sea level rise and land subsidence), two risk points are identified at the exposed Manzala lagoon barrier.

Conclusions

This means that not all the coastal plains of the Nile Delta and Alexandria are vulnerable to accelerated sea level rise at the same level due to the high variability of the land topography. The continued and increasing rate of sea level rise will primarily affect the vulnerable areas on the lower coastal plain of the Nile Delta but not the entire plain. The benefits gained from understanding the interactions between the rate of sea level rise, subsidence, topographic variability, structural faulting, and seismic activity can help reduce the uncertainties associated with local sea level rise projections, thus contributing to more effective coastal management of the study area, including adaptation strategies that would help minimize the negative consequences on socioeconomics and ecological impacts.

Coastal Protection Measures, Case Study (Mediterranean Zone, Egypt)
https://doi.org/10.1007/s11852-015-0389-5

Abstract-Summary

The coastal zone suffers from sedimentation, accretion, and pollution problems as well as the side effects of climate change. Major efforts have been exerted to manage coastal erosion problems and to restore coastal capacity to protect housing, infrastructure, and cultivated land. These problems were encountered with various types of hard structures, but most of these methods' responses were limited due to a lack of evaluation of the entire ecological situation. This encouraged coastal engineers to think about new types of environmentally friendly structures that work better with the ecological situation. There are several efforts to apply these new technologies to protect the coastal zone as well as the environment by taking into consideration the effect of climate change. These new approaches in coastal protection are multi-use, environmentally friendly, easy to modify and maintain, and efficient from an economic perspective. Previous efforts in solving coastal problems in Egypt are analyzed and discussed, taking into consideration the experience of similar cases worldwide. The environmentally friendly coastal structure is more suitable to solve most of our coastal problems by saving our ecosystem and reducing the protection cost.

Introduction

Coastal erosion is a global problem (Cai et al. 2009). With this new approach, the concept of coastal protection is changed from hindering natural forces to building with nature (Van Rijn 2004). This new approach, which is building with nature, requires knowledge of the exact behavior of the coastal zone and hence addresses the main reasons for the problem in order to choose its suitable protective structure. Egypt, as one of the pioneers in coastal protection, should start work with these new concepts to protect its coastal environment and increase its economic value. This contribution presents an overview of the various available methods for shore stabilization and beach erosion control and highlights a new approach in coastal protection to recommend a proper solution for Egyptian coastal problems. Protection of the coast and the shore against the forces of waves, currents, storm surge, and flood can be performed in many ways. There are two kinds of protective measures for controlling coastal erosion: structural measures and nonstructural measures (Cai et al. 2009). Nonstructural measures include land-use controls, setting warning lines such as the coastal setback line and coastal construction control line to protect the coast from improper construction, and the prohibition of unreasonable sand mining and reclamation.

Types of coastal protection measures and their usage

Submerged breakwaters can reduce beach erosion and protect coastal structures by dissipating a significant amount of wave energy (Cho et al. 2001). Low crested and submerged structures such as detached breakwaters and artificial reefs are becoming very common coastal protection measures used alone or in combination with artificial sand nourishment (Pilarczyk 2003). The relatively new innovative coastal approach is to use artificial reef structures called "Reef Balls" as submerged breakwaters, providing both wave attenuation for shoreline erosion abatement and artificial reef structures for habitat enhancement. Protection—the impacts of sea level rise are controlled by soft (e.g., beach nourishment) or hard (e.g., dike construction) engineering, reducing human impacts in the coastal zone that would be impacted without protection. The adaptation measures that were identified to address the impact of climate change on coastal zone areas include beach nourishment, construction of groins and breakwaters, tightening legal regulations, integrated coastal zone management, and introducing changes in land use (Batisha 2012).

Proposed coastal protection for the Egyptian coasts

It is difficult to obtain a proper technique for all kinds of coastal problems or preferred from an environmental standpoint (Iskander 2010). The international highway could be elevated to 5 m above sea level with "a wall on the side facing the sea"; sand dunes should be protected and restored to function as natural walls; all development in the Delta ought to be subject to integrated management and planning to ensure that no new structures are put in the wrong place (Frihy et al. 2010a, 2010b). Proposals for solving problems in the Nile Delta coastal zone: First, there is an erosion problem due to the lack of sediment supply from the River Nile. This problem can be overcome by dividing the beaches into small cells by using headlands. The flood problems of the low land and sabkha can be partially stabilized by using sand dunes with vegetation to support the outside face. Constructed jetties with sand bypass from the upstream side to the downstream side can also solve this problem.

Conclusion

During this investigation, the aim was to assess each type of coastal protection by identifying its advantages and disadvantages to suggest a suitable protection method for the Egyptian coasts. It is shown from the above cases that most traditional ways of protection, which were used around the world and our Egyptian coast, have side effects on the environment and beach morphology. The findings of this study suggest that using innovative techniques with soft techniques (beach nourishment, sand dune stabilization) can significantly affect the restoration of beaches without affecting the environment or the habitat of aquatic organisms. The second major finding, particularly on the Egyptian coast, is that many coastal problems can be solved by reaching the natural situation before the building of the Aswan high dam. Further work needs to be done to identify whether innovative techniques are suitable for the Egyptian coast.

Climate Change Impacts, Vulnerabilities, and Adaptation Measures for Egypt's Nile Delta
https://doi.org/10.1007/s41748-018-0047-9

Abstract-Summary

A significant number of institutions, research centers, universities, and governments have funded projects in addition to work done by independent scholars and assessors studying this phenomenon, in particular, to identify vulnerability, mitigation, and adaptation against associated risks. Egypt is among the international community and has participated in numerous studies, research activities, conferences, seminars, and meetings attempting to address climate change and its associated risks. The aim of this paper is to review current and previous projects, technical reports, and pilot studies concerning risk assessments, mitigation, and adaptation strategies for climate change in Egypt. This, in turn, will aid in decision-making regarding future funding and establishing research related to climate change in Egypt.

Introduction

Climate change-associated risks could pose serious threats to Egypt. Expected significant fluctuations in the River Nile flow, sea level rise (SLR), and temperature increases due to climate change are the direct threats that climate change poses to Egypt. For more than two decades, Egypt has been conducting climate change studies. This includes the work by El Raey et al. (1995) on the vulnerability of coastal resources; Strzepek et al. (1995) on the vulnerability of agriculture to changes in climate, water supply, and coastal resources; El-Shaer et al. (1996); and Eid et al. (2006) on the impacts of changes in climate on agriculture and crop production. Its focus is to cover a broad range of the research conducted on climate change in Egypt, particularly concerning the Nile Delta since it is the most populated area and contributes to more than 50% of Egypt's agricultural land (FAO 2015). This paper covers three main subjects, namely, (1) climate change impacts, (2) vulnerability, and (3) adaptation and mitigation studies.

Climate Change Impacts

The Intergovernmental Panel on Climate Change (IPCC) indicates that the primary cause of SLR is the thermal expansion of ocean waters due to glacier melting resulting from increased greenhouse gas emissions (IPCC 1990). As CO_2 concentrations increase, temperature increases, leading to greater water demand as water loss increases due to evaporation and crop water requirements. Water is a limiting factor in Egypt, as the Nile flow is expected to fluctuate at all scenarios for temperature increase (Agrawala et al. 2004). Egypt's current development plans will magnify the increasing pressure on land and water resources. Climate change impacts on agriculture threaten the country's food security given that Egypt produces only 60% of its food and only 40% of its grain requirements (Racha 2015). Climatic change, overpopulation, rapid growth rate, urbanization, industrial development, and irrigation intensification increase water demand and magnify the vulnerability of agriculture in Egypt.

Climate Change Vulnerability

The Nile Delta is vulnerable to the impacts of storm surges and SLR resulting from climate change, particularly the relatively low-elevation areas in the Delta (Batisha 2012). Sea level rise and intensified storm surges could cause severe impacts on the lower Nile Delta, parts of Alexandria and Port Said in terms of inundation, soil salination of the low-lying lands, and erosion of coastal barriers (Eldeberky and Hünicke 2015). Areas vulnerable to inundation could be categorized based on topography, subsidence rate, SLR, coastal barriers, protection, and groundwater levels. Eldeberky and Hünicke (2015) identified the lowland of southeast Alexandria, the southern zone of Port Said, Burullus and Manzala Lakes barriers, and Ras El bar shoreline as more vulnerable to SLR and coastal inundation. Land subsidence also threatens the Nile Delta lowlands, which could increase the impacts of SLR. Alexandria, Beheira, and Damietta are very vulnerable to climate change impacts.

Climate Change Adaptation and Mitigation

In 1999, Egypt highlighted its required adaptation plans and actions for climate impacts and vulnerabilities in the Initial National Communication and then updated it in the Second National Communication in 2010 (Arab Republic of Egypt 2010). The content of the Initial National Communication was a result of research conducted between 1995 and 1999 that included agricultural and water resource vulnerability, adaptation to SLR, and evaluation of the technologies implemented in the action plans. The focus of the Egypt plan was to create a greenhouse gas emissions inventory for different sectors, prioritize policies and measures for emissions reduction, and conduct a socioeconomic impact assessment of these measures; however, adaptation needs were not addressed in the Initial National Communication (Arab Republic of Egypt 1999). Egypt's NEEDS for climate change was built on the work of the country's second national communication, identifying specific adaptation needs for the agricultural sector and coastal zones.

Conclusion

Climate change would increase temperature and reduce rainfall amounts. The significant climate change impacts on Egypt seem to be the fluctuations of the water flows in the River Nile and inundation of the coastal areas due to SLR. Significant uncertainty remains regarding exactly how climate change will affect the flow of the River Nile. Uncertainty surrounding the Nile flow indicates that any discussions on the impacts of climate change in Egypt need to take into account three scenarios: lower Nile River flow, higher flow, and unchanged high variable flow (Link and others 137). Growing high-yielding cultivars in suitable agro-climatological regions and sowing at the right time will increase crop production, thereby reducing the impact of climate change (El-Shaer and others 128).

A Holistic Evaluation of Human-Induced LULCC and Shoreline Dynamics of El-Burullus Lagoon Through Remote Sensing Techniques

https://doi.org/10.1007/s41062-020-00331-w

Abstract-Summary

Monitoring changes occur in land use/land cover carried out using Landsat and SPOT images over the years (1984, 2001, 2012, and 2017). The results of the change detection revealed that the lake area has decreased by 7.23% due to the continuous increase in the urban area around the lake. It is concluded that the maximum rate during the period (2001–2012) is 46.7 m/year due to the construction of hard protection on the west side, which resulted in downdrift erosion (eastern side) with a maximum rate of 32 m/year. The maximum erosion rate was 38.3 m/year on the eastern side between 2012 and 2017, while the maximum erosion rate on the western side was 48.65 m/year.

Introduction

Coastal areas are subject to various complex natural processes that always trigger long- and short-term changes: shoreline retreat, movement of sediments, degradation of water quality, and coastal growth. Historical recorded wave data from 1977 to 2010 were analyzed to examine the threat of climate change on the wave conditions of the Nile Delta coast (Iskander 2013). Wave energy is estimated to increase by approximately 20% in high storm events and decrease by approximately 1% in normal conditions over the next 50 years in the El-Burullus coastal region. Accurate and updated shoreline change positions are important because they are used in various coastal management studies, such as investigating hot spot areas along coastal zones, forecasting potential shoreline positions, coastal protection, and sustainable management of coastal resources (Davidson et al. 2010; Louati et al. 2014; Szmytkiewicz et al. 2000). The prime objective of the current study is to analyze the evolution of the coastal area of El-Burullus using a supervised classification technique, thus detecting changes during the duration (1984–2017).

Materials and methods

Preprocessing the image is very important to be more suitable for different processes in remote sensing science (Akhter 2006). In supervised classification, to indicate a certain class, the user selects certain pixels of the image with the same properties. The supervised classification was applied to the four images using the maximum likelihood approach. The water reflection is almost equal to zero in the reflectivity infrared bands, and a single band image can be used to extract the shoreline. Another technique is to use the supervised classification of imagery by the selection of training sites as mentioned earlier; the shoreline is typically mapped from remote sensing data. The analysis results of data from supervised classification and digitized shorelines for different dates (1984, 2001, 2012, and 2017) are shown in vector layers after converting classified images from raster to vector.

Results and analysis

Since the shoreline of the study area is too long, the whole area was divided into two subareas to show accretion/erosion patterns: western and eastern parts. The maximum erosion rate on the eastern side of the lake between 1984 and 2001 was 20 m/year, whereas on the western side, the maximum accretion rate was 14 m/year. The maximum accretion rate occurred during the period (2001–2012), with a rate of 46.7 m/year due to the construction of hard structures that hindered and trapped the transported sedimentation driven by longshore drift, resulting in instability of the sediment budget in the area. It caused accretion on the western side and erosion on the downdrift (eastern side) with a maximum rate of 32 m/year. During 2012–2017, the maximum erosion rate on the eastern side was 38.3 m/year, while the maximum accretion on the western side was 48.65 m/year.

Conclusions

It is noticeable that erosion still occurs on both the western and eastern sides of the lake. It is concluded that the maximum rate of erosion between 1984 and 2001 was 20 m/year on the eastern side of the lake, while the maximum rate of accretion was 14 m/year on the western side. The maximum accretion rate occurred during the period (2001–2012) with a rate of 46.7 m/year due to the construction of hard protection on the western side resulting in erosion downdrift (eastern side) with a maximum rate of 32 m/year. During 2012–2017, the maximum erosion rate on the eastern side was 38.3 m/year, while the maximum accretion on the western side was 48.65 m/year.

Natural and Anthropogenic Influences in the Northeastern Coast of the Nile Delta, Egypt

https://doi.org/10.1007/s00254-008-1434-6

Abstract-Summary

Landsat enhanced thematic mapper imagery (ETM) from 2002 and aerial photography from 1955, combined with published charts and field observations, were used to interpret coastal changes in the zone between the Kitchener drain and Damietta spit in the northeastern Nile Delta, previously recognized as a vulnerable zone to the effects of any sea level rise resulting from global warming. The interpretation resulted in the recognition of several changes in nine identified geomorphological land types: beach and coastal flat, coastal dunes, agricultural deltaic land, sabkhas, fish farms,

Manzala lagoon, saltpans, marshes, and urban centers. Reclamation of vast areas of the coastal dunes and of Manzala lagoon added approximately 420 km^2 to the agricultural deltaic land. Approximately 48 km^2 of backshore flats, marshes, saltpans, and Manzala lagoon have been converted to productive fish farms. The main urban centers have expanded; nearly 12.1 km^2 have been added to their areas, and new urban centers (Damietta harbor and the New Damietta city) with a total area reach of ~35.3 km^2 have been constructed at the expense of vast areas of Manzala lagoon, coastal dunes, and backshore flats. Shoreline changes have been determined from beach profile surveys (1990–2000), and a comparison of 1955 aerial photographs and the ETM satellite image of 2002 reveals alongshore patterns of erosion versus accretion.

Introduction

Based on aerial photographs taken in 1955 and field observations, Frihy et al. (1988) classified the delta morphology into six major units: the nearshore zone up to 7 m water depth, beach and coastal flats, coastal accretion sand ridges, carbonate ridges, coastal dunes, and coastal lakes. Like many other deltas, the coastal zone of the Nile Delta is currently undergoing extensive changes caused by both natural and anthropogenic influences. Natural factors affecting changes in the Nile Delta coast include sediment supply, coastal processes, tectonic activities, land subsidence, climatic fluctuations, and sea level rise. With regard to beach erosion, the coastal zone of the Nile Delta faces other environmental issues, including saltwater intrusion, irrational land use, water pollution, and deterioration of natural resources and habitats. The aims of this paper are (1) to monitor and update the environmental changes in the coastal zone between the Damietta spit and Kitchener drain in the northeastern Nile Delta and (2) to determine the short-term shoreline changes in the context of the morphodynamic behavior of the delta.

Materials and methods

Short-term shoreline changes along the study area were determined from the available data from beach profile surveys conducted from 1990 to 2000. Change in the measured distance between the baseline point and the shoreline position (y) over the date of profile survey (t) permit accurate determination of the mean annual rates of shoreline changes (m/year) using the least-squares regression analysis. This annual rate of shoreline changes combined with coastal processes is incorporated to analyze the morphodynamic behavior of the study area. In a separate effort, the two rectified maps of 1955 and 2002 were compared to assess the rate of shoreline change during a 47-year time span. The two shorelines were superimposed, and the rate of change between them along a fixed baseline point was calculated using AutoCAD.

Results and discussion

Further east, accretion exists behind the detached breakwaters of Ras El Bar (maximum of 15 m/year) between profiles P27 and P38 and reverses to mild erosion with a rate of −3 m/year at the downdrift side. In this subcell, sand eroded from the relict Burullus–Baltim promontory and coastal dunes is transported downcoast to the east, resulting in accretion along the Gamasa embayment. Introduction of these structures has blocked or retarded, in most cases, the eastward transport of beach sand and in turn has induced marked shoreline change, particularly downdrift erosion or updrift accretion and too much extent forming sandy tombolos on the lee side of the shore-parallel breakwaters. This reversed longshore current is largely responsible for transporting sediments from the eroded zones west of the Damietta promontory tip and from Ras El Bar beaches toward the Gamasa sink.

Conclusions

The total area of cultivated land has enlarged; reclamation of vast areas of the coastal dunes and of Manzala lagoon added approximately 420 km^2 to the agricultural land. The main urban centers have extended at the expense of Manzala lagoon, agricultural deltaic land, coastal dunes, and backshore flats; approximately 12.1 km^2 has been added to their area. The analysis of shoreline positions established from the beach profile survey along the coastline of the study area spanning the years 1990–2000 and comparison of the aerial photograph of 1955 and ETM satellite image of 2002 reveal alongshore patterns of erosion versus accretion. In the Burullus subcell, sands eroded from the relict Burullus–Baltim promontory and coastal dunes, including the most western coast of the study area, are transported downcoast to the east by the prevailing littoral currents and currents generated by the east Mediterranean gyre, resulting in accretion along the Gamasa embayment.

Economic Evaluation of Using Marine Dredged Material for Erosion Control Along the Northeast Coast of the Nile Delta, Egypt

https://doi.org/10.1007/s12517-016-2660-y

Abstract-Summary

The study also provides an economic evaluation for the shoreline management alternatives selected to mitigate the effects of coastal erosion in two pilot eroding areas (namely, A and B) located near Damietta Harbor. For soft nourishment by dredged sand, other types of coastal engineering measures, which are often used in erosion management areas, were also evaluated as alternatives for erosion control and mitigation solutions. Analysis of alternatives has also been supported by other criteria to select the cost-effective and environmentally acceptable option to protect the eroding pilot areas. The final selection of the best viable alternative indicates that the procedure of beach nourishment is the most appropriate form for protection area A, while a combination of groins and sand nourishment is more relevant for area B. In any case, material dredged from the navigation approach of Damietta Harbor should be utilized as a borrow material in the nourishment schemes and excluding the use of terrestrial sources.

Introduction

In the absence of sediment supply to the Mediterranean coast off the Nile Delta, the continued action of coastal processes combined with the effect of accelerated sea level rise acts to induce beach erosion (UNDP 1978; Fanos et al. 1991; Deabes 2003; Frihy and El Sayed 2013). Taking this into consideration, the use of dredged material from Damietta Harbor in the nourishment of nearby eroding beaches can effectively help combat accelerated erosion and management of Nile Delta shoreline resources. The reuse of dredged sediments would also cost less than borrowing beach nourishment material from offshore sources, which can also induce unfavorable environmental aspects, such as accelerating erosion rates along adjacent coasts. It aims to assess the feasibility of the maintenance-dredged material in Damietta Harbor for nourishment of nearby eroded beaches and remediation of accelerating erosion along the Nile Delta coast.

Dredging operations in Damietta Harbor

The eastern one is approximately 500 m long and extends to approximately the 3 m water depth contour. The navigation channel is approximately 11.4 km long, 200–300 m wide, and has a maximum water depth of approximately 15 m. Since 1986, the harbor has been experiencing sedimentation problems, particularly in the 3-km inner approach channel, and therefore requires frequent dredging to foster safe navigation depth and innocent passage of ships (Frihy et al. 2002). It is planned to finalize the processes of mineral separation in the black sand placer plant at Burullus (under construction), ~70 km west of the sedimentation basin. Three sedimentation basins have already been constructed in that area to receive the expected slurry, with a maximum depth of 3 m and a total capacity of 750,000 m^3.

Hydrodynamic processes and erosion rates of the study shores

The beach sand has an average mean grain size of 0.128 mm and sorting of 0.823ϕ (or standard deviation). To the immediate east of area B, the shoreline is developed into a long and narrow shore-parallel sand spit. Beach sand has an average mean grain size of 0.117ϕ and a mean sorting of 0.643ϕ. The backshore spans the western part of the Manzala lagoon and is composed of swamps and fish farms; the latter is characterized by linear and rectangular features. Data obtained were processed to calculate the annual rate of beach changes by applying the Digital Shoreline Analysis System (DSAS), an ArcGIS tool developed by the US Geological Survey (Thieler et al. 2005). The system has been set up to calculate rates of beach change at a series of cross-shore transacts at an interval of 100 m alongshore.

Current shore protection scheme

The results of applying the morphodynamic numerical Delft3-D model by Bahgat and Ramadan (2015) are consulted in the present study to facilitate the most appropriate nourishment scheme and to simulate expected variations in the seabed of the beach surf zones of the two study eroding areas A and B. The input data of the model-included waves in 20 m water depth, tidal variations, bathymetry, shoreline position, grain size distribution of native, and borrow sediment. From this figure, for scenario 1, the beach width at areas A and B would increase by approximately 0.102 and 0.297 km^2, respectively, with a shoreline advance of approximately 50 and 40 m after 1 year following nourishment, respectively, compared to the original shore without nourishment. Four potential shoreline protection alternatives were selected for the mitigation of erosion effects in areas A and B. These alternatives include conventional hard engineered structures and nourishment by borrowing material from dredging activities (with or without heavy mineral separation) or from other terrestrial sources.

Cost–benefit analysis

Benefits are mostly related to the monetary values of land use/land cover, a unit backing the coastline of study areas A and B. Geomorphologically, this land coverage pattern differs in the study areas and includes beach and strandplain in areas A and B, as well as fishing resources in wildlife and aquaculture lagoon areas in area B. The present land use/land cover was identified using Google Earth images and eventually validated by field observations. The land use/land cover of the two backing areas (A and B) are converted into monetary benefits by application of the contingent valuation method. Contingent valuation is used to estimate the total economic value of the identified land use/land cover backing the study areas.

The cash outflow, however, is the total benefits (revenues) due to the protection of areas A and B; this flow contains revenues of new and existing beach, strandplain, swamp, fish farm, and lake.

Extraction of black sand from dredging material

Frihy et al. (2015) indicated that sediments dredged from Damietta Harbor and some other navigation waterways along the Nile Delta coast are relatively rich in black sand and therefore can be exploited as a promising source for valuable economic heavy minerals. The average total heavy mineral concentrations in sediments dredged from the Damietta Harbor channel were found to be 2.9%, with predominantly higher concentrations (weight per kilogram) of magnetite (1.298%), ilmenite (1.266%), hematite (0.005%), leucoxene (0.030%), garnet (0.031%), zircon (0.133%), and rutile (0.103%). The results indicate higher concentrations relative to the study of Frihy and others (150), in which the total heavy mineral content is 6.5% and consists of magnetite (0.38%), ilmenite (0.57%), rutile (0.16%), leucoxene (0.13%), garnet (0.13%), zircon (0.21%), and monazite (0.007%).

Comparison of erosion control alternatives

End results indicate that the procedure of beach nourishment is the most appropriate form for protection area A, whereas a combination of groins and sand nourishment is more relevant for area B. In both options, material dredged from the navigation approach of the Damietta Harbor should be utilized as a borrow material in the nourishment operations. This comprises a loss of EGP 8.4 billion due to unexploitation of the dredging material for economic heavy minerals and an additional EGP 4.0 billion due to the cost of dredging and dumping of potential nourishment sand in offshore areas. Based on all of the aforementioned results, the present study strongly recommends exploitation of the dredging material as a source of nourishment sand for combating ongoing erosion in areas A and B. The use of borrowed sand from inland sources should be avoided because of its negative environmental impacts and high total cost.

Conclusions and recommendations

The results of compatibility analysis indicate that sediment dredged from Damietta Harbor in the northwestern Nile Delta is technically feasible for beach nourishment of both eroding areas A and B and for separation of economic heavy minerals. Among the alternative protection measures selected, the option of beach nourishment was found to be the most appropriate form for protecting area A, and a combination of groins and sand nourishment was most appropriate for area B; both are cost-effective and environmentally sound measures. In both options, material dredged from the navigation approach of the Damietta Harbor should be utilized as a borrow material in the nourishment operations and exploited for nourishment sand and economic heavy minerals. Governmental institutions in Egypt, such as SPA and Damietta Port Authority, should take serious actions toward exploiting the dredging material for environmental and economic uses instead of their dumping without any recycling or exploitation.

Evaluation of Future Land-Use Planning Initiatives for the Shoreline Stability of Egypt's Northern Nile Delta

https://doi.org/10.1007/s12517-017-2893-4

Abstract-Summary

An analysis of beach morphodynamics and relative vulnerability to erosion processes and sea inundation within the Nile Delta's lower coastal plain was performed as a basis for evaluation of future planning initiatives undertaken by the National Centre for Planning State Land Uses (NCPSLU). The appropriateness of shores fronting the planned land uses is assessed in the present study by applying the coastal vulnerability index (CVI) approach, which depends primarily on a variety of variables that affect the beach stability of the proposed plans. The results revealed that not all of the examined delta coastlines are equally vulnerable to beach erosion (ranging from relatively low to highly susceptible). Future planning of unprotected highly vulnerable and low-elevation subsiding zones, such as the Manzala lagoon barrier, requires effective mitigation measures.

Introduction

Emphasis here is on the evaluation of the planned Nile Delta's coastal margin with regard to the effect of ongoing factors responsible for controlling its coastal vulnerability, including hydrodynamic processes, rising sea level, land subsidence, saline water intrusion into the delta's aquifer, diminished water flow, and sediment discharge to the shore. The susceptibility or vulnerability of the new land-use areas planned in the Nile Delta's plain margin is evaluated herein based on the potential effects of erosion induced by the combined influence of diverse factors: (1) morphodynamic changes of the examined coast, analyzed here based on shoreline change rates versus wave attacks and (2) possible sea inundation due to increased eustatic sea level rise, resulting from increased polar glacial ice melt and thermal expansion of seawater associated with global warming (Hansen et al. 2016, 151). The Nile Delta coast is presently subjected to submergence as a partial response to a rise in eustatic sea level, plus the effects of land lowering (subsidence).

Methodology

The Abuodha and Woodroffe (2006) CVI is calculated as the square root of the product of the ranked variables divided by the total number of variables, where a1 = dune height, a2 = barrier type, a3 = beach type, a4 = RSL change, a5 = shoreline erosion or accretion, a6 = mean tidal range, and a7 = mean wave height. Consulting these sources, the input variables derived herein are dune height = 0 to 20 m (a1), barrier types = main land (a2), beach types = dissipative (a3), relative rates of sea level change = 2 to 4.9 mm/year (a4), rate of shoreline changes = $\geq$ +2 m/year (accretion) and $\leq$ − 2m/year (erosion) (a5), mean tidal range = 0.4 m (a6), and mean wave height = 1.2 m (a7). For the CVI calculations, the stability of the coastline position as shoreline change rates is calculated here using the coastal surveys performed in the last 15 years.

Results and discussion

The relevancy or usefulness of the coastlines fronting the newly planned land uses is discussed here below based on the interaction between their vulnerability to the effect of erosion and sea inundation supported by information on the "Nile delta littoral cells" and the existence of nodal points. Sand eroding from the upcoast of these structures is transported to the east by the prevailing unidirectional NW wave-induced longshore current, where it is locally blocked by these structures, causing downdrift erosion. Low CVI values (to <4) occur primarily along concave shores, such as bays, embayments, saddles, and updrift sides of major engineering structures, which have a higher tendency toward sand deposition under the effect of low wave energy convergence. The fish farm is partially fronted by a very high vulnerability shore that might need appropriate protection measures to confront the effects of land lowering (subsidence) and low-lying nature of this sector.

Summary and conclusions

The focus of the present study is to assess the suitability levels of the land-use plans formulated by the NCPSLU to develop unused stretches in the lower Nile coastal plain through the concept of coastal vulnerability assessment. In marked contrast, eroding shores along the unprotected coastlines and those at downdrifts of engineering structures, with high to very high CVI = 6 to > 8, are most likely threatened for land-use planning unless effective protective structures are built. In highly dynamic subsiding low-elevation, delta plain margins such as the Nile Delta, coastal vulnerability and viability of future land-use plans should be assessed first to ensure conformity with physical factors interplaying to produce erosion and inundation, including coastal morphodynamics and RSLR. Vulnerability assessment should also consider the analysis of shore morphodynamics and current and future protective measures, significantly influencing coastal stability.

Effects of Shoreline and Bedrock Irregularities on the Morphodynamics of the Alexandria Coast Littoral Cell, Egypt

https://doi.org/10.1007/s00367-004-0178-x

Abstract-Summary

Beach-nearshore profiles combined with beach and surficial sediment samples were analyzed in conjunction with wave, current, littoral drift, and sea level data to determine the effect of bedrock on morphodynamic processes within the littoral zone of Alexandria on the Mediterranean coast of Egypt. The compartmented nature of this cell acts together with the rough irregularity of the rocky seafloor to trap a thin veneer of sediment (<3 m thick), showing proportional mixing between two sedimentary provinces. A modern fine-grained sediment facies consisting of mixed carbonate/siliciclastic sand flanks most of the nearshore zone down to a depth of 8–10 m. Beyond this depth, considered to be the depth of closure, a relict late Pleistocene to mid-Holocene coarse-grained facies composed of biogenic carbonate sand is found. Along a short section of the coastline (km 3–6), the coarser sediment also occupies the nearshore zone. The present shoreline orientation, headlands, and rough, irregular rocky seabed are reflected in the erosion/accretion pattern, sediment characteristics, and reversibility of longshore currents and littoral drift. The physical properties of the bedrock (degree of induration), the accelerating sea level rise during the Holocene and human influences, the modern morphology of the coast, the erosional seabed features in the nearshore zone, and the texture of seabed sediments are all controlled by the original geometry of the coast, which consisted of an elevated subaerial ridge.

Introduction

The ancient city of Alexandria is located immediately west of the Nile Delta and was founded by Alexander the Great in 332–331 BC. The coast west of Alexandria is characterized by straight, carbonate-sand beaches of variable width. The origin of these ridges ranges from marine, such as offshore bars and beach deposits, to aeolian (Fourtau 1893; Shukri et al. 1956; Said et al. 1956; Butzer 1960; Lindell et al. 1991). The purpose of this study is to evaluate coastal processes, seasonal factors, and human interventions influencing morphodynamic changes in the littoral zone of Alexandria, particularly the role of beach and bedrock materials in governing the morphology of the beach and littoral zone sediment budget as well as the prevailing driving forces. Based on the morphodynamic nature of the littoral zone and the elevation of the upland area, the vulnerability of the Alexandria waterfront to the predicted rise in sea level is evaluated.

Environmental setting

The Mediterranean Sea around Alexandria shows a typical semidiurnal microtidal regime. Long-term daily gauge records indicate maximum and minimum astronomical tidal water levels of +33.8 and −15.9 cm, respectively, giving a maximum tidal range of 49.7 cm (Frihy 1992). Storm surges can raise the sea level by approximately 40 cm, causing overtopping of the beach and the Corniche (Hamed and El Gindy 1988).

Engineering structures

Due to the sediment deficiency in the Alexandria littoral cell, beaches were protected as early as 1934 to prevent further erosion and to provide new recreational sandy stretches. Breakwaters have been constructed to protect El Diekhila Harbor, Western Harbor, Eastern Harbor, and Abu Quir Harbor. Nourishment projects have been effective in protecting and improving amenities on eroding beaches. Between 1999 and 2002, two detached breakwaters were constructed to reduce coastal erosion at Cleopatra Beach.

Materials and methods

Relative percentages of four sand-sized components were calculated from point counts of >400 grains per sample (method according to Frihy and Stanley 1988). Longshore current measurements were obtained inside the surf zone at water depths ranging from 1.2 to 1.5 m by tracking the movement of a buoy and measuring the time it took to travel a distance of 20 m in the longshore direction. The modified N-line model of Perlin and Dean (1983) was applied to determine wave directions (exposures) inducing longshore transport along the coastline of the study area. The results obtained enabled us to diagrammatically determine the wave angle, which would change the net longshore sediment transport direction from NE or SW. In defining wave exposure, we identified those components in the wave rose of Alexandria that were able to induce longshore transport on both sides of the shore-normal line.

Results

The statistical relationships, the spatial distribution patterns of grain size parameters (mean grain size and grain sorting), and the carbonate contents of beach and seabed sediments are used in this study to interpret sedimentary provinces and morphodynamics. The relative percentages of heavy minerals are higher in the fine-grained sediment group (6%) and very low in the coarse-grained sediment group (0.9%). The data show that a mean grain size of 0.45 mm clearly delimits two groups of data points, corresponding to a fine- and a coarse-grained sand population. The average wave direction-height frequency distribution reveals that wave heights between 0.5 and 1.0 m dominate over those with heights of > 1–2 m. The wave data show that low-swell waves prevail during spring and summer, and wave heights rarely exceed 1.2 m. The N, NNW, and NW waves are important in inducing morphological changes because of their long duration, particularly in winter.

Discussion

Previous studies have assumed that the sediment grain size of the littoral zone reverts gradually to the original distribution after a certain period following beach nourishment (Krumbein and James 1965). Previous studies have used distinct changes in sediment characteristics (color, texture, heavy minerals, and benthic faunas) along shore-normal profiles as geological indicators to delineate the depth of closure of the seaward depth limits at which significant wave energy levels interact with a sand bed (Hallermeier 1981). The long-term exposure of the relict coarse-grained sand facies indicates that they are not influenced by the nearshore fine-grained sediment dispersed by seaward-directed currents. This situation differs from that recorded at the Nile Delta and indicates that southwesterly littoral currents at Alexandria are strong enough to transport sediment along the coast.

Conclusions

The littoral cell of Alexandria is unusual in that it is longshore transport-dominated, has a small sand budget, and receives little sand from freshly weathered rock, artificial nourished sand, and palimpsest offshore sediment. These sediments show seasonal cyclicity induced by changes in wave climate. The irregular bedrock of the Alexandria littoral cell is blanketed by two sand facies exhibiting proportional mixing of carbonate-siliciclastic sands controlled by sediment sources, morphologic features, hydrodynamic processes, and sea level rise. The coarse-grained sediments existing offshore can be used as a source of "borrow" material for beach nourishment.

Assessment of Natural Coastal Hazards at the Alexandria/Nile Delta Interface, Egypt

https://doi.org/10.1007/s12665-020-09329-0

Abstract-Summary

Three vulnerable spots [2.3–2.5 m above mean sea level (msl)] are identified at Alexandria, specifically at the Eastern Harbor, Lauran, and Mandara-Montaza beaches. Three other low-lying spot areas are distinguished at Idku resort beach (1.0–1.5 m above msl) and on both sides of the western and eastern flanks of the Rosetta Nile promontory (1.0–1.2 m above msl). Detailed analysis of spatial and cross-shore transects generated from the near shore/land elevation map marks contrasting elevations that vary from low-lying (− 3 m below msl) to high-elevation carbonate ridges parallel to the shore (~20 m above msl). The high-elevation coastal ridges underlying Alexandria's seafront—maximum elevation of 12 m above msl—in addition to other protection elements are acting together as a natural quasibarrier to mitigate sea flooding that may in turn affect the historical low-lying depressions located east and southeast of the city.

Introduction

On a nationwide scale, relative SLR at the Nile Delta margin is induced from the combined processes of eustatic (world) sea level in the Mediterranean, the effects of land subsidence triggered by neotectonics and ongoing compaction of the Holocene soft mud section (to ~50 m) (Stanley 1990; Stanley and Clemente 2017). For land subsidence, beach erosion is also considered a major issue despite multiple hard stabilization or mitigation measures built at Alexandria and the Nile Delta coast. With the exclusion of wave runup and sea inundation, a number of articles have discussed the coastal vulnerability of the Nile Delta. These studies provide a general conclusion that the Nile Delta margins and, to some extent, the city of Alexandria are highly vulnerable to SLR due to their low-lying elevation (El Sayed 1991; Sestini 1992; El Raey et al. 1999; Frihy 2003; El Raey 2010; Batisha 2012; Hassaan and Abdrabo 2013). The present study differs greatly from previous studies in which the focus is on potential coastal hazards at the Alexandria/Northwestern Nile Delta interface based on in situ measured seafront elevation and contiguous seawater depth.

Methodology

To reduce uncertainties, a high-resolution surface elevation map of the study area is created from closely spaced cross-shore elevation measurements surveyed across the waterfront of the study area between 500 and 800 m width. Geodetic benchmark leveling data were measured at 44,396 points in 2017 by the Surveying Research Institutes via a Total Station and a Differential Global Positioning System (DGPS) with a relative accuracy of < 1 m distance and $\pm$ 10 cm elevation. The in situ measurements are complemented with inland elevations generated from a series of recent topographic maps. Depth soundings (24,483 data points) are measured through 55 cross-shore beach profile data points using an echo-sounder with a vertical accuracy of ± 10 cm in conjunction with a DGPS. The slope of beach profiles up to— 6 m depth from msl is calculated to determine wave runup elevation at 50 predefined stations.

Case (3): Irregular wave runup design for rough, impermeable slopes

Z_{op} the deep-water Iribarren number based on peak period T_p in meters. Annual RSLR is estimated from the yearly averaged tide-gauge data measured at Alexandria Western Harbor (1944–2006) and the Rosetta estuary (1982–2018). Some of the data points of these time series have been previously reported by Frihy et al. (2010a, 2010b). Coastal stability and dynamics of the study area are investigated by estimating rates of shoreline change in open sea beaches with no harbors during a 10-year period from 2008 to 2018. Shoreline positions are measured along the length of the study coast using a portable DGPS connected with a marine PC.

Results and discussion

At Alexandria, sea flooding of 50–60 m distance is observed along the ~2.3–2.5 m lowest points at the Eastern Harbor, Lauran, and Mandara-Montaza Chornish road. Visual comparison indicates that the Alexandria seafront seems to be secure with respect to sea flooding, as its seafront surface is generally elevated higher than 2.5 m above msl. In marked contrast to the above and as might be expected, the low-lying Nile Delta seafront, which is elevated slightly above 1.0 m from msl, would be automatically vulnerable to sea flooding or inundation in the case of a 1.0 m rise in sea level. Along the western coast of Alexandria from Sidi Krair to El Dekhila Harbor, 15.4 km long, low rates of erosion (−2.5 m/year), and accretion (3.0 m/year) prevail.

Conclusion

A high-resolution contour map of the Alexandria/Northwestern Nile Delta interface, generated from in situ measurements of waterfront surveys and topographic maps, supplemented by nearshore depth soundings and wave data, is incorporated with other physical criteria to assess major natural hazards. These spots are positioned at Alexandria (Eastern Harbor, Lauran, and Mandara-Montaza beaches) and the Nile Delta (Idku resort beach and on both sides of the Rosetta Nile Delta promontory). Based on the vulnerability criteria discussed earlier, the identified spots in the Nile Delta are mostly at high risk compared to those in Alexandria. The three vulnerable spots in Alexandria are threatened by wave runup and sea inundation, but in the case of two of them (the Eastern Harbor and Lauran), the risk shoreline is very narrow and could be mitigated more effortlessly.

Erosion and Accretion Rates and Their Associated Sediment Characteristics Along the Ras El Bar Coast, Northeast Nile Delta, Egypt
https://doi.org/10.1007/s00254-006-0447-2

Abstract-Summary

Beach profile data, covering the coast of Ras El Bar, northeast Nile Delta, collected during the years from 1990 to 2002 combined with Landsat images for the area and sedimentological investigation have been used to identify beach and nearshore seafloor sediment changes. Along the coast of Ras El Bar, two accretion sectors and one erosion sector have been recognized. Both sectors are characterized by a growing shoreline with maximum rates of ~15 and 10 m/year, respectively. They have maximum nearshore seafloor accretion rates of ~18 and 22 cm/year, respectively. The erosion sector is located east of Damietta Port and has a maximum rate of shoreline retreat of ~ −10 m/year. Erosion of its nearshore seafloor is indicated, recording a maximum rate of ~ −20 cm/year. The two accretionary sectors are characterized by the dominance of moderately sorted fine sands in their shore area, which change seaward into less sorted very fine sands.

Introduction

Approximately 450 BC, Herodotus recorded that the Nile Delta was built up by the alluvium discharged in the Mediterranean Sea by seven branches. It grew in the sea approximately 5–8 km from 1800 to 1900 (Sestini 1976); a reversal toward an erosion phase began in approximately 1900 due to the reduction in the Nile sediment load. Since the construction of the High Aswan Dam in 1964, the discharge of Nile River sediment has stopped, and the rates of erosion and shoreline retreat have accelerated (Frihy and Khafagy 1991; Inman and Jenkins 1984). Changes along the Nile Delta coast have attracted the attention of many scientists. Local and regional interpretations of short-term volumetric sediment changes along the delta have been made by Manohar (1976) and El Fishawi and Badr (1989).

The study area

Erosion along the coast of Ras El Bar resort has been mitigated by the construction of a series of coastal engineering structures, including jetties, groins, seawalls, and detached breakwaters. In 1941, a jetty was constructed on the western side of the Damietta Nile branch to reduce the deposition of drifted sediment. A second jetty was constructed on the eastern side of the Damietta Nile outlet in 1976 to reduce siltation and shoaling at the Damietta branch outlet. At the southern end of the western jetty, a concrete seawall was constructed in 1963 to stop erosion of the shore. The western breakwater was constructed to be approximately 1,400 m in length and is almost parallel to the navigation channel; it extends to the 7.00 m depth contour [measured from the lowest low water level (LLWL)].

Methods of study

The database for the present study was extracted from the survey of Ras El Bar beach profiles carried out by the Coastal Research Institute (CoRI) biannually through an intensive program to monitor the changes in the Nile Delta coast since 1972. This variable provides data on the bottom changes over the time of the profile survey. The data for each profile are arranged in a three-dimensional array, where h is the water depth relative to the mean sea level, y is the shoreline position relative to a benchmark, and t is the date of the profile survey. These data permit the determination of the mean annual rates of bottom changes (cm/year) and shoreline changes (m/year). The sediment volume change (m^3/m) over the period of the profile survey was determined from the profile survey data using a computer program adopted by CoRI. Using a grab sampler, bottom samples were obtained jointly with the profile survey.

Coastal processes affecting the northeastern Nile Delta

Ras El Bar is affected by seasonable changeable waves (Fanos and others 177; Fanos 178). In spring, the maximum wave height reaches 1.16 m, the average is 0.4 m, and the predominant wave direction is NW. In winter, the maximum wave height reaches 4.2 m with an average of 0.51 m, and the predominant wave direction is NNW. The overall maximum wave height is 4.25 m; the average wave height and period are 0.50 m and 6.5 s, respectively.

Results and discussion

In the first sector of accretion, west of Damietta harbor, sediment volume steadily increased along the sector coincident with the accretion of the sea bottom and shore; maximum rates of 658, 768, and 1,196 m^3/m in the years from 1990 to 2000, 2001, and 2002, respectively, were recorded. In the second accretion sector, which is located in the shadow of the detached breakwater system, sediment volume change has maximum rates of 605, 443.3 and 448.5 m^3/m in the years from 1990 to 2000, 2001, and 2002, respectively. On the east side of the sector, sediment volume changed from 483.3 and 280.3 m^3/m in the years from 1990 to 2000 and 2001, respectively, to −85.3

m^3/m in the years from 1990 to 2002, indicating reversals of accretion to erosion as a consequence of sediment deficiency on the downdrift side of the breakwater system.

Summary and conclusion

The change in beach and nearshore sediments in the area extending from the west of Damietta Port eastward to the outlet of the Damietta Nile branch during the years from 1990 to 2002 indicated two sectors of accretion and one of erosion. At a distance of 600 m from the shore, beyond the breaker zone, the maximum accretion rate of the seafloor for the first sector is 10 cm/year and for the second sector is 2 cm/year. Both sectors are characterized by the dominance of moderately sorted fine sands in their shore area, which change seaward into relatively less sorted very fine sands. The erosion sector is located east of Damietta Port and has a maximum rate of shoreline retreat of approximately 10 m/year. Alongshore and nearshore distributions of heavy minerals in the study area indicate total heavy enrichment in the sector of erosion on the beach and seafloor of the breaker zone.

Vulnerability and Fate of a Coastal Sand Dune Complex, Rosetta-Idku, Northwestern Nile Delta, Egypt

https://doi.org/10.1007/s00254-007-0912-6

Abstract-Summary

The types, distribution, and origin of recent sand dunes between Rosetta and Idku in the western sector of the Nile Delta, Egypt, were investigated. Sand samples from the dunes, beach, and seafloor were studied for grain size distribution and mineralogical composition. Most of the dunes in the study area have been subjected to deterioration and removal due to the construction of buildings and the International Coastal Highway. The blown sand grains accumulated to form a belt of coastal sand dunes of original longitudinal and crescentic forms. The study area is considered vulnerable to the impacts of climate change and the expected rise in sea level. Protection and preservation of the remaining dunes in the study area are vital requirements for shore protection.

Introduction

Many studies have dealt with the coastal sand dunes in the northern part of the Nile Delta. Barakat and Imam (1976) identified ancient indurated sand dunes in the district of Gamasa, northern Nile Delta. They classified the dunes in the area into two main types: young active sand dunes and old stabilized dunes. Abu Diab (2001) and Barakat (2004) evaluated Nile Delta coastal sand dunes as a source for economic heavy minerals. El-Banna (2004) described the coastal sand dunes in the area between El-Burullus and Gamasa and discussed their nature and human impacts. He highlighted the role of sand dunes as a dam protecting the central part of the Nile Delta shoreline and cultivated land on the backshore. This project was initiated in 2002 to study the coastal sand dunes along the Nile Delta.

Material and methods of study

Samples were collected for grain size and mineral analyses. Area, samples were collected at 100 m intervals along 22 beach profiles perpendicular to the coast and extending seaward to approximately 6 m water depth. The study area, which includes the western sector of the Nile Delta coast, has an arid climate with a mean annual rainfall of 7.4 in. (18.8 cm). The Nile Delta coast is affected by a wave regime that varies with season (Fanos et al. 1995; Fanos 1996). In spring, maximum wave heights can reach 1.16 m with an average of 0.4 m; the predominant wave direction is NW. In winter, the maximum wave height reaches 4.2 m with an average of 0.5 m, and the predominant wave direction is NNW. The overall maximum wave height is 4.25 m, and the average wave height and period are 0.51 m and 6.5 s, respectively.

Data analysis and results

The mean grain size values (Mz) for all sand dune samples range from 0.11 to 0.25 mm. Heavy mineral species similar to those identified in the dune sands have been identified in beach and seafloor sediments of the study area. The overall constituents of heavy minerals in the beach, nearshore seafloor, and dune sands are those assemblages that characterize the Nile deposits (Shukri 1950). According to O'Keefee (1978), grasses and other obstacles on the berms trap wind-blown sand from the beach; thus, the berms are increased in width and height to form coastal dunes. Incipient foredunes; the small sand mounds formed around bushes on the beach berm parallel to the shoreline manifest the beginning formation of a new generation of coastal dunes. Mitigation plans for the maintenance, protection, and enhancement of the coastal sand dunes in the study area should be considered as soon as possible.

Summary and conclusion

The main types of sand dunes in the study area are linear, barchan, and incipient foreshore dunes. The wide coastal flats existing to the north and northwest have an important role in the development of the dunes in the study area. Accumulation of sand on the coastal plain east of the promontory of the former Canopic Nile branch formed the dunes in the study area. A major area of the coastal sand dunes between Rosetta city and El Geddia village has been urbanized. Foreshore sand dunes are important to coastal protection. The manufactured destruction of these coastal sand dunes leads to negative impacts on the area and its environment. Developing, repairing, and protecting the remaining dunes in the study area are vital requirements for strengthening shoreline defenses against beach erosion and are effective in mitigating the possible consequences of future sea level rise.

Evolution of Modern Nile Delta Promontories: Development of Accretionary Features During Shoreline Retreat

https://doi.org/10.1007/s00254-004-1103-3

Abstract-Summary

The active accretion features that have developed along the modern Nile Delta promontories during shoreline retreat are analyzed using topographic maps, remote imagery, ground, and hydrographic surveys, together providing 15 time-slice maps (1922–2000) at Rosetta and 14 time-slice maps (1909–2000) at Damietta. Both the Rosetta and Damietta inlets are associated with submerged mouth bars that accumulated prior to the damming of the Nile but that continue to contribute to local sedimentation problems, particularly at Rosetta. The development of the active accretion features along the Nile promontories reflects a combination of factors, including sediment availability, transport pathways from source areas, a decrease in the magnitude of Nile flood discharges, and the impact of protective structures at the river mouths.

Introduction

Analysis of historical maps spanning the nineteenth and twentieth centuries indicates that during the nineteenth century, the promontories progressively extended into the Mediterranean Sea in response to sediments delivered to the coast by the Rosetta and Damietta branches of the Nile. Following the cessation of sand delivery to the coast, the continued action of waves and longshore currents has resulted in a major reorientation of the coastline (UNESCO/UNDP 187; Blodget and others 95; Inman and others 188), with maximum rates of erosion occurring along the outer margins of the Nile promontories. Previous studies of the morphology of the Nile Delta coast have tended to focus on shoreline change with relatively little emphasis on local depositional features, such as spits and sandbars. The purpose of this paper is to evaluate temporal changes in the occurrence of sandy accretion features located at the outlets of the Rosetta and Damietta branches of the Nile River and to discuss several factors that have contributed to their formation.

Data sources and methodology

The compiled data provide 15 time-slice maps for the Rosetta promontory, spanning the period 1922–2000, and 13 time-slice maps for the Damietta promontory, spanning the period 1909–2000. The long-term rate of coastal retreat was estimated based on the position of the promontory tip in each successive time slice, referenced to a fixed baseline. The baselines for the Rosetta and Damietta promontories were placed landward of the contemporary shoreline at latitudes of 31° 27′ 00″ and 31° 25′ 40″, respectively, and the measured displacements were corrected for local shoreline orientation. The resulting displacements as a function of time were subjected to least-squares regression to estimate the average retreat rates of the Rosetta and Damietta promontory tips during the past century. Large-scale patterns of erosion and accretion along the entire promontories were also assessed based on aerial photographs from 1955, ground surveys, and satellite images from 2000.

Results

The SPOT image of 1986 shows a very narrow spit growing from the western flank of the promontory and associated with a detached mouth bar or shoal directly in front of the inlet, as indicated by breaking waves. To the Rosetta spits, a submerged river mouth bar or shoal formed immediately seaward of the inlet. Unlike the Rosetta inlet, sedimentation at the Damietta inlet is not spit-related but mainly takes the form of submerged river mouth bars. Time-slice maps and beach surveys of accretionary spit and bar features at the Rosetta and Damietta mouths indicate that they are very dynamic in time and in space. Sand eroded from the promontory tip is transported both to the west and east at the Rosetta mouth, producing a double spit feature.

Summary and conclusions

At the Rosetta promontory, two short spits developed at the river mouth following the construction of seawalls. The spit has persisted since its formation but is presently exhibiting some evidence of erosion, which reflects the depletion of its principal sediment source at the outer margin of the promontory as a consequence of coastal protection works. The development of the Damietta spit and the appearance and disappearance of the Rosetta spits are controlled by several diverse factors, including sediment availability, wave and longshore currents, variability in the Nile branch discharges, and the construction of protective structures at the river inlets. A similar situation appears to now be developing at the Damietta promontory, with some erosion evident along the seaward margin of the spit, most likely a consequence of a diminished updrift sediment supply due to the construction of the seawall in 1995.

Pre- and Postbeach Responses to Engineering Hard Structures Using Landsat Time Series in the Northwestern Part of the Nile Delta, Egypt
https://doi.org/10.1007/s11852-008-0013-z

Abstract-Summary

Analyses have been undertaken to examine shoreline positions established from remote sensing data along the northwestern part of the Nile Delta from Abu Qir Bay to the Gamasa embayment (~143 km length). Digital shoreline analysis software was used to calculate the annual rate of beach changes at 1,432 cross-shore transects prior to (1972–1990) and after protection (1993–2006). Maximum shoreline retreat occurs along the Rosetta promontory (−138.52 m/year) and along the central bulge of the delta at Burullus headland (−6.07 m/year). Areas of shoreline accretion exist within saddles or embayments between the promontories at west Abu Qir Bay (20.04 m/year), Abu Khashaba saddle (16.17 m/year), and Gamasa embayment (20.68 m/year). These rates of change have been significantly altered by the construction of intensive shoreline protective structures built from 1990 to combat areas of rapid erosion at both the Rosetta promontory and Burullus–Baltim headland, ~15 km in total length.

Introduction

These factors include a reduction in Nile Delta discharge and sediment loads due to the construction of dams and barrages across the Nile River, combined with the action of wave-induced longshore currents. Wave-induced currents have redistributed accreted sediments, leading to sedimentation problems in the navigation channels of the delta harbors and the estuaries of the Rosetta and Damietta channels of the Nile. In these subcells, sand eroded from a promontory (Rosetta and Damietta promontories) or headland (Burullus headland) is transported to the east and is either deposited naturally in the next embayment or is trapped by an artificial structure, resulting in shoreline accretion. Coastal dynamic processes (Manohar 1981; Khafagy and Manohar 1979; Fanos 1986; El Wany et al. 1988; Nafaa et al. 1991) and sediment transport (Inman and Jenkins 1984; Frihy et al. 1998b; El Sayed et al. 2005) have disclosed the relation between wave-induced longshore currents and processes of erosion and accretion along the delta coastline. A number of coastal protection works, such as jetties, groins, seawalls, and breakwaters, have been constructed along the Nile Delta promontories to combat beach erosion and to reduce inlet shoaling.

Methodology

Two techniques are used to estimate the rate of shoreline retreat. Technique, estimate rate of shoreline change based on a time series of Landsat satellite data the by using Digital Shoreline Analysis System (DSAS) program. To extract the shoreline position, an image threshold was formed for band 4 (0.8–1.1 μm) for MSS or band 7 (2.0–2.35 μm) for TM/ETM+ (shortwave infrared) on each date to form a binary image or image mask (zero value for water and one value for land). To ensure the waterline mapping accuracy in the case of MSS image data, a 3×3 edge enhancement filter was used to sharpen the boundary between the water and land classes. The data measured from each profile are then used to estimate the mean annual rate of shoreline change (meters per year) employing linear regression techniques.

Results and discussion

Prior to protection (1972–1990), the highest rate of erosion is centered on the Rosetta Promontory (−138.52 m/year), as waves and longshore currents transport sand away from the Rosetta branch of the Nile, which no longer supplies new sand to the adjacent beaches. Further east, the sandy barrier separating the Burullus lagoon and the Mediterranean has experienced lower erosion rates of −15.13 m/year before protection with detached breakwaters. To the east, along the unprotected beaches and sand dunes (5.7 km long), erosion varies between −0.98 and −2.77 m/year. Following sheltering of the Burullus–Baltim bulge headland by intensive hard structures, this accretion that originally prevailed in the preprotection period along the Gamasa embayment reverses to erosion that systematically increases eastward, where it attained a maximum of −10.34 m/year at approximately 0.5 km east of this drain.

Conclusions

The large-scale rate of beach changes along the study area is now being altered as a response to hard structures built within the littoral cells, particularly at Rosetta promontory and Baltim resort beach. The erosion along the Burullus–Baltim headland is replaced by accretion (13.26 m/year) due to the construction of a series of shore-parallel breakwaters and groins; originally, this erosion was approximately −5 m/year prior to protection of this area. These points position areas of transport from erosion to deposition or vice versa that result from the orientation changes of the shoreline. Other nodal points are observed east of the Rosetta seawall because of the construction of five groins to mitigate downdrift erosion in this area. The alongshore variability in shoreline change rates is closely consistent with areas experiencing variable energy levels that are undergoing erosion and/or accretion.

Morphodynamic Implications for Shoreline Management of the Western Mediterranean Sector of Egypt

https://doi.org/10.1007/s00254-008-1595-3

Abstract-Summary

Although the western Mediterranean coast of Egypt between Sallum and Alexandria, ~550 km long, has maintained considerable equilibrium throughout history, developers have built traditional protective structures in an effort to form sheltered recreational beaches without taking into consideration their geomorphologic characteristics, coastal processes, and harmful impacts on the coastal environment and human safety. The results deduced will have practical applications for shoreline management initiatives regarding sustained sites suitable for future beachfront development, such as safe swimming conditions, sport facilities, water intakes, and sheltered areas for vessels. The benefits realized by understanding morphodynamic processes would enhance our awareness of the significance of the role of western coast morphodynamics in supporting sustainable development via shoreline management. On a national scale, the results reached could provide a reliable database for information that can be used in establishing a sustainable shoreline management plan, which is, in turn, an essential part when implementing an Integrated Coastal Zone Management Plan for this region of attraction.

Introduction

The shore-parallel ridges are entirely composed of Pleistocene limestone ridges, which represent the source of carbonate oolitic sand for the beaches of the western coast and Alexandria (Hilmy 1951). This implies that these beaches are much more dynamic in morphologic changes under varying wave conditions and are associated with hazardous rip currents, particularly in gently sloping areas. They mostly have a relatively steep slope and narrow surf zone where waves break close to the shore and develop directly into an intense swash that runs up and down the beach surface. Beach erosion is now widespread along much of the western coast of Alexandria due to anthropogenic influences rather than natural processes. Assessing the degree of vulnerability of significant shoreline features to wave energies is presented in this study to help enhance tourism activities and support shoreline management plans for sustainability, such as scenic areas, swimming, water sports, beaches, and natural harbors.

Methodology

High-resolution satellite images are combined with field observations to characterize the morphologic configuration of the study coastline. To document the seabed morphologies, 18 beach-nearshore profiles were surveyed by the Egyptian Coastal Research Institute (CRI 2000) at nine selected stations along the range of the study area, with two profiles at each site. The profile's basic survey was conducted using a Nikon Total Station model (power set 3010), in addition to graded staff and a leveling instrument. Marine surveys were conducted using a rubber boat equipped with a computerized differential geographic position system (DGPS) with a relative accuracy of <1 m and horizontal accuracy ±5 m and a Navi-Sound 205 Echo-sounder with ±10 cm accuracy.

Results and discussion

In significant contrast, littoral cell 5 between Shaqiq headland (Hamra) and El Dekhiela is a long sandy beach with small or no clear headlands; however, part of the shoreline is interrupted by hard structures built by resorts for recreation and swimming entertaining activities. The direction of the longshore current is deducted from the analysis of the relationships between shoreline orientations and prevalent incident wave directions at main coastal segments: long beach, headland, and headland leeside. Based on our understanding of morphodynamic processes acting in littoral cell 3, the entire coastline between Sallum and El Dekhiela can be classified into five distinctive morphotypical features according to the degree of wave exposure or shelter, ranging from high energy in open coastal areas to low energy in sheltered basins. Since the acute angles of wave crests along the shoreline are very large, the sediment transport rate or littoral drift is relatively high, estimated to be 0.23×10^6 m^3 per year at Marakia beach (CRI 1993) and between 0.03 and 0.16 m^3 per year at Alamien-Marina Resort (Delft Hydraulics 2000).

Summary and conclusions

The objective of this study has been to provide decision-makers and planners with sufficient insight into the coastal system of the western Mediterranean coast of Egypt to select potential sectors valid for conducting new development projects such as safe swimming conditions, sport facilities, water intakes, or sheltered areas for vessel harboring. Such analyses have been made to evaluate the relationship between shoreline configuration and prevailing coastal processes in an attempt to provide a framework for determining the degree of coastal vulnerability to wave exposure. degree of natural protection. Based on the degree of natural protection in relation to wave exposure, the western coast of Egypt can be categorized into different morphotypical features: long-smooth shorelines, headland tips, headland leesides, bays and embayments, pocket beaches, and naturally sheltered areas. It is recommended, however, to use different protective measures, such as submerged hard structures combined with sand nourishment and "perched beaches", to protect semisheltered embayments, bays, and long coastlines.

Environmental Impacts of Baltim and Ras El Bar Shore-Parallel Breakwater Systems on the Nile Delta Littoral Zone, Egypt
https://doi.org/10.1007/s00254-003-0886-y

Abstract-Summary

This study evaluates the impacts resulting from the construction of two large-scale detached breakwater systems on the Nile Delta coast of Egypt at Baltim and Ras El Bar beaches (~18.3 km shoreline length). The beach and nearshore sedimentation (erosion/accretion patterns) and grain texture of seabed sediment in the study areas have been substantially disrupted due to the interruption of longshore transport by the shore-parallel detached breakwaters. The preconstruction beach erosion at Baltim (−5 m/year) and at Ras El Bar (−6 m/year) has been replaced by the formation of sand tombolo (35 m/year) and salient (9 m/year), respectively. Beach erosion has been substantially increased on the downdrift sides of these protective systems, being −20 m/year at Baltim and −9 m/year at Ras El Bar. Further seaward, the two protective systems at Baltim and Ras El Bar have accumulated seabed sand at maximum rates of 30 and 20 cm/year and are associated with downdrift erosion of −45 and −20 cm/year, respectively. Strong gyres and eddies formed in the breakwater gaps have drastically affected swimmers and subsequently caused a significant number of drownings each summer, averaging 35 and 67 victims/year at Baltim and Ras El Bar beaches, respectively. Losses of sediments, with a maximum of −803 m^3/m, are indicated downcoast of these structures between breakwaters #6 and 9.

Introduction

The dramatic erosion along some stretches in the delta has been mitigated by the construction of a series of coastal engineering structures, which started extensively during the last decade, including jetties, groins, seawalls, and detached breakwaters. Structures, two major detached breakwater systems, have been constructed to combat beach erosion along the Nile Delta and span ~10 km in total. The breakwater systems of Baltim were constructed from land-based access and not from the sea. The main objective of this paper is to assess the coastal impacts induced by constructing the two breakwater systems at the beaches of Baltim and Ras El Bar. This study differs from previous investigations in that we concentrate herein on the two recently built detached breakwater systems in relation to their physical and human impacts versus coastal processes and their design implications.

Detached breakwater design parameters and beach changes

Detached breakwaters have been widely used as effective measures to combat beach erosion, particularly along tideless coasts such as the Mediterranean. Along sandy coasts, the construction of detached breakwaters mostly affects the shoreline by creating either a salient or tombolo in its shadow area. The formation and development of salient or tombolo is a function of diverse mechanisms related to offshore breakwater design, including breakwater crest height (H), effect of the offshore distance of the breakwater (X), and the length (B), and gab (G) of the breakwater (Wiggel 1964; Rosen and Vajda 1982; Herbich 1991; Dean et al. 1997). To minimize the formation of tombolo, detached breakwaters should be constructed to a design that will allow waves to partially dissipate on the breakwater while at the same time allowing some wave energy to pass through to prevent the formation of the tombolo.

Materials and methods

Annual profiles surveyed between 1990 and 2002 spanning the two study areas were analyzed in this study to determine the rate of seabed changes, sediment volume, and grain size distribution of seabed samples in relation to coastal processes. The measured vertical distance of the water depth (H) between the seabed and the sea surface at MSL provides a reliable means of monitoring bottom changes over the period of profile collection. Bottom changes were determined every 100 m along each of the profile lines, measuring from the shoreline. The data from each profile are arranged in a 2-D graph, where H is the water depth relative to mean sea level and X is the date of survey. The measured shoreline displacement from the fixed baseline (Y) is used to monitor the shoreline changes over the time of profile collection.

Results and discussion

This erosion is higher than that which occurred before construction of the breakwaters all along Baltim beach, which originally was −5 m/year (Frihy et al. 2003). On the opposite sides of this area, this erosion is reversed to accretion to the west at the updrift side of Damietta harbor (maximum of 1,200 m^3/m) and in the shadow area of the breakwater system (maximum of 605 m^3/m). The maximum longshore transport rate is calculated in the eroding area between Damietta harbor and breakwater #8 (449 × 10^3 m^3/year) and decreases in accreting areas behind the breakwaters (~100 × 10^3 m^3/year maximum). These relationships result from the processes of selective grain sorting as the waves and longshore currents first erode the sand from the beach face, transport the sand alongshore, and finally deposit it in areas of accretion (Frihy and Komar 1991).

Summary and conclusions

The present study was undertaken to determine whether the implemented detached breakwater systems that were built along the beaches of the Baltim and Ras El Bar resorts (18.3 km long) have affected the sedimentation patterns of littoral cells and swimmers using adjacent bathing water. A comparison of the long-term background erosion rates with those determined in the last 12-year monitoring period suggests that the breakwaters caused an additional erosion rate above the background erosion by approximately 300 and 50% at Baltim and Ras El Bar beaches, respectively. At the two study areas, the detached structures are of higher relative crest elevations (emerged breakwaters) and were not sufficient to allow so much water to flow over the breakwaters and avoid tombolo formation. Because the Baltim breakwaters were built at a relatively shorter offshore distance than those of Ras El Bar (X/B < ~0.6 to 0.8), tombolo has been well developed broadly where it is replaced by salient formation at Ras El Bar (X/B > ~2.0).

Responses of Ras El Bar Seafloor Characteristics to Protective Engineering Structures, Nile Delta, Egypt
https://doi.org/10.1007/s00254-005-0065-4

Abstract-Summary

The bathymetry of the seafloor in front of the Ras El Bar coast, the characteristics of the seafloor sediments, and the effects of protective structures on seafloor bathymetry were examined. Seafloor depths in front of the Ras El Bar seawall ranged between 2 and 4 m. These depths, coupled with the wave climate, resulted in waves breaking directly onto the wall and exerting high, dynamic pressures on the area at the structure's toe. West of the seawall, the constructed breakwater system has affected the bathymetry of the seafloor in the area. The gaps between the breakwater units have attained deep depths and steep slopes. Scours and steep slopes adjacent to the head of the breakwater units have been recorded. Seaward of the breakwater system, deeper depths, and gentle slopes characterize the seafloor.

Introduction

The Nile Delta is subjected to dramatic erosion due to coastal processes and delta subsidence, reduction in the Nile discharge, and sediment load to the Nile promontory mouths because of the construction of water control structures along the Nile (UNESCO/UNDP 1978; Stanley and Warne 1998). Erosion along the Ras El Bar resort coast has been mitigated by the construction of a series of coastal engineering structures, including jetties, groins, seawalls, and detached breakwaters. The first protective structure in the area was a jetty constructed in 1941 on the western side of the Damietta Nile branch to reduce the deposition of drifted sediment. Due to progressive erosion of the coast at the southern end of the west jetty, a concrete seawall was constructed in 1963. Previous studies on the effect of these protective structures on the Nile Delta coast have been limited to shoreline changes.

Coastal processes affecting the Ras El Bar area

In spring, the maximum wave height reaches 1.16 m, with an average of 0.4 m at a predominant NW wave direction. In winter, the maximum wave height reaches 4.2 m with an average of 0.5 m and a predominant NNW wave direction. The overall maximum wave height is 4.25 m; the average wave height and period are 0.51 m and 6.5 s, respectively. The predominant wave directions generate westward-flowing longshore currents.

Methods of study

The database for the present study was extracted from the survey of Ras El Bar beach profiles carried out by the Coastal Research Institute (CORI) biannually through an intensive program to monitor the changes in the Nile Delta coast since 1972. The beach profile survey took place biannually during calm conditions in September–October and April–May. Using a grab sampler, bottom samples were obtained jointly with the profile survey. Samples were collected at 100 m intervals along each profile, yielding 150 samples. Grain size analyses of each sample were performed using standard sieving and pipette techniques, and a size distribution pattern was obtained for each sample.

Results and discussion

Two adjacent sectors of different characteristics are shown along the Ras El Bar coast; the first is in front of the Ras El Bar seawall, and the second is behind the breakwater system. Offshore–onshore water currents and sediment movement toward the northeast have been inferred from the distribution pattern of the contour lines.

Bathymetry of the seafloor in front of the Ras El Bar seawall

Placement of a rock blanket with adequate bedding material seaward from the toe of the structure will cause shoaling of the seafloor and wave breaking before striking the structure (US Army Crops of Engineers 1984). Preventing breaking of the waves directly on the wall will minimize the scour near its base and result in a more stable structure. The width and stone size of the toe apron required to prevent scouring of the seafloor in front of the structure are related to several factors. A detailed study of scouring at the natural bottom and near existing structures should be conducted, and a model study should be considered before determining a final design for the toe apron width and stone size. Based on Eckert (1983), the minimum toe apron is inadequate for protection against wave scouring if the following two conditions exist. If the waves break directly on the toe apron, the minimum quarry stone weight will be inadequate regardless of its slope.

Bathymetry of the seafloor in the breakwater system area

According to El-Banna and Hasaneen (1999), this sector is a zone of accretion because the breakwater system acts as a littoral sediment trap. The detached breakwater system has caused the accumulation and growth of sediment from their leeward shoreline. The configuration and values of contour lines in the bathymetric map indicate that sediment growth has continued under the water level, forming submerged spits. This dramatic change in bathymetry of the gap areas between the breakwater units is the result of current and eddy effects (El-Banna and Hasaneen 2002).

Bottom sediment characteristics

Heavy minerals increase in the seafloor sediments near the seawall and decrease in the shadow of the breakwater system. Dominant heavy minerals in the sediment are opaques (magnetite and ilmenite), pyroxenes (augite), amphiboles (hornblende), and epidotes with small amounts of garnet, monazite, zircon, rutile, and tourmaline. Opaques, garnet, monazite, zircon, rutile, and tourmaline are relatively high in the sediments near the seawall where the seafloor is eroded and scoured, in agreement with Frihy and Komar (1993), who stated that heavy minerals of relatively higher density tend to remain in the area of erosion.

Summary and conclusions

The bathymetric map of the Ras El Bar area shows relatively deep depths and steep seafloor in front of the seawall base. Monitoring of beach profiles in front of the structure indicates that it steepens with time and forms scour troughs in front of the structure base. Seafloor depth in front of the structure combined with wave climate in the area results in waves breaking directly on the wall and exerting high, dynamic pressures at the point where the wave crests hit the structure. Beach profiles are changed, and scour troughs are formed in front of the structure. Garnet, zircon, monazite, rutile, and tourmaline are present in relatively higher quantities in the sediments near the seawall base.

Forecasting Shoreline Changes Along the Egyptian Nile Delta Coast Using a Landsat Image Series and Geographic Information System
https://doi.org/10.1007/s10661-021-09192-x

Abstract-Summary

It became a highly destructive Delta due to the lack of sediment discharge, climate change, subsidence, and coastal processes (e.g., wind, waves, tides, and littoral currents). Many coastal structures have been erected to stop or mitigate coastal problems in the study area. The linear regression ratio (LRR) and end-point rate (EPR) were used with Digital Shoreline Analysis System (DSAS) software to determine the rates of beach changes; we then forecasted future shoreline changes. The accuracy of the model's results was checked using the ground field measurements of several studies. The value of the uncertainty is low (approximately half a pixel) along the shorelines without coastal protection. This study aimed to forecast future beach evolution to 2041 to evaluate its sensitivity and facilitate proposals for coastal protection for human safety and habitats if coastal processes and climate change continue to worsen with time.

Introduction

Several protection works have been established to minimize coastal changes along the coast of the study area, including seawalls, groins, dikes, jetties, and breakwaters. These coastal works have completely stopped further erosion of the Rosetta and Damietta headlands and have caused beach changes along other parts of the Egyptian Mediterranean coast. Several previous studies have found changes in the pattern of erosion and sedimentation on the coast of the Nile Delta. Many studies (Dewidar and Frihy 2007, 2010; Frihy and Komar 1993; Frihy et al. 2010a, 2010b; Sestini 1989; Stanley 1988a) have determined that erosion rates in the Rosetta headland are the highest along the delta coast (−96 m/year). The objectives of this study included further monitoring of the Egyptian Nile Delta for 46 years and updating its coastal geomorphology to determine appropriate protection measures in the long term. This approach will help decision-makers put forward a future plan to protect the Egyptian Nile Delta coast if coastal processes and climate change continue to worsen with time.

Study area

It includes six major cities—Alexandria and Rosetta to the west, Burullus and Baltim in the middle, and Damietta and Port Said to the east. The area includes five local governorates—from the west Alexandria, El Beheira, and Kafr El Shiekh, and Damietta and Port Said to the east. Approximately 19 million people live in the coastal Nile Delta area. The maximum temperature is 40 °C in August, and the minimum is 10 °C in winter months.

Physical settings of the Nile Delta

Geomorphologically, the Nile Delta consists of beaches, coastal dunes, deserts, and lakes. Three coastal lakes are located on the Egyptian Nile Delta coast. A study by Fanos and others (219) found that the prevailing wind direction on the coast of the Egyptian Nile Delta during the year (50–60%) is from the north and northwest direction, while the northeastern winds account for 10–15%. Wave action along the Egyptian Delta coast is seasonal in nature, responding to changing wind systems over the water where the waves are generated (Gad et al. 2013). Frihy et al. (2010a, 2010b) produced wave rose diagrams for different areas along the Nile Delta coast. Iskander et al. (2007) measured wave records along the Nile Delta coasts from 1985 to 2010 and recorded an increasing trend in sea wave height from 2.6 to 2.9 cm/year.

Protection structures along the Nile Delta

Several coastal protection structures, including jetties, groins, seawalls, and breakwaters, were built along the promontories of the Egyptian Nile Delta to combat beach changes and reduce inlet siltation. These included five groins constructed in 2003 with lengths ranging between 400 and 500 m seaward, spaced 800–900 m apart. In 2005, construction began on another ten short groins (80–150 m long) at distances of 500–600 m on the lee side of the western seawall (Frihy et al. 2010a, 2010b). Farther east, a concrete 600-m-long jetty was built in 1950 to protect the eroding beach downstream of the Burullus inlet. These breakwaters were constructed at a depth between 3 and 4 m in the active surf zone. At the Damietta promontory, two jetties were constructed to protect the entrance of the Damietta inlet from silting between 1941 and 1963 (Dewidar and Frihy 2010).

Materials and methods

Image processing software, such as Erdas Imagine, v. 15, helps improve the accuracy of feature extraction. AutoSync software in Erdas Imagine was used for all Landsat image data. To ensure waterline mapping accuracy in the case of MSS image data, 3 × 3 edge enhancement filters were used to sharpen the boundary between water and land classes. Horizontal and vertical Sobel filters were used for each unsupervised classified image on each date to enhance edge detection. The filtered images for each date were converted to vector layers using the raster-to-vector module. The updated rate of shoreline change is used to forecast successive time steps until another survey date is reached, and new data are processed again in the model. All shoreline data were extracted from a single source (Landsat image series).

Results and discussion

Sediments are being deposited in the Idku sink at a rate of 6 m/year. The eroded sediments from the Damietta headland drift into two parts—the Ras El Bar Resort to the east and the Damietta Spit to the west. Nine detached breakwaters were built along the coast of the Ras El Bar resort to protect swimmers and infrastructure. Farther toward the Port Said breakwater, there is additional deposition of sediment eroded from the Damietta headland. Application of this model in the study area affirmed, with a high degree of confidence, the behavior of beach erosion and accumulation if the conditions of natural coastal processes and protection work are not changed. This model showed disruptions in the pattern of erosion and sedimentation due to the installation of new groins, especially in the eastern part of the Rosetta Branch.

Conclusions

Erosion and sedimentation rates were calculated over 46 years for a 233-km stretch along the study area. The three headlands of the Nile River, the shape of the Delta coast, suffer from erosion and accretion. In the long term, climate change will cause catastrophic situations in low-lying areas around the world, particularly along the Egyptian Nile Delta coast. The application of this model succeeded in clarifying general patterns of erosion and sedimentation along the coast of the Egyptian Nile Delta, which should encourage decision-makers to develop strategic plans and protection measures.

References

Abd-El Monsef H, Smith SE, Darwish K (2015) Impacts of the Aswan High Dam after 50 Years. Water Resour Manage 29:1873–1885. https://doi.org/10.1007/s11269-015-0916-z

Abdel-Aziz T, Dawod G, Ebaid H (2020) DEMs and reliable sea level rise risk monitoring in Nile Delta, Egypt. Discov Sustain 1(1). https://doi.org/10.1007/s43621-020-00006-7

Abdrabo MA, Hassaan MA (2015) An integrated framework for urban resilience to climate change—case study: sea level rise impacts on the Nile Delta coastal urban areas. Urban Clim 14:554–565

Abou Samra RM (2017) The use of cartographic modeling to assess the impacts of coastal flooding: a case study of Port Said Governorate, Egypt. Environ Monit Assess 189:435

Abou Samra RM, El-Gammal M, Al-Mutairi N, Alsahli MM, Ibrahim MS (2021) GIS-based approach to estimate sea level rise impacts on Damietta coast Egypt. Arab J Geosci 14:429. https://doi.org/10.1007/s12517-021-06810-3

Abu Diab AA (2001) Sedimentological studied and evaluation of the coastal sand dunes between El Geddia and Idku. Egypt M. Sc. Thesis, Fac Sci, Mansoura University, El Mansoura

Abuodha PAO, Woodroffe CD (2006) Assessing vulnerability of coast to climate change: a review of approaches and their application to the Australian coast. University of Wollongong, Faculty of Science Papers, Research Online. Available on line at: http://ro.uow.edu.au/

Abutaleb K, Mohammed A, Ahmed M (2018) Climate change impacts, vulnerabilities and adaptation measures for Egypt's Nile delta. Earth Syst Environ. https://doi.org/10.1007/s41748-018-0047-9

Agrawala S, Annett M, El-Raey M, Conway D, van Aalst M, Hagenstad M, Smith J (2004) Development and climate change in Egypt: focus on coastal resources and the Nile. OECD, Paris. http://www.oecd.org/dataoecd/57/4/33330510.pdf

Akhter M (2006) Remote sensing for developing an operational monitoring scheme for the Sundarban Reserved Forest, Bangladesh. Unpublished Doctoral Thesis. Technische Universität Dresde, Dresden

Arab Republic of Egypt (1999) Initial national communication on climate change. Prepared for The United Nations Framework Convention on Climate Change (UNFCCC)

Bagheri M, Ibrahim ZZ, Mansor SB, Abd Manaf L, Badarulzaman N, Vaghefi N (2019) Shoreline change analysis and erosion prediction using historical data of Kuala Terengganu, Malaysia. Environ Earth Sci 78(15):477

Bahgat M, Ramadan Kh (2015) Dredged material placement to support sustainable shoreline management in the Nile Delta, internal report, Ramadan, Hydraulic Research Institute, National Water Research Center, Internal Report, 62 pp

Barakat MG (2004) Sedimentological studied and evaluation of some black sand deposits on the northern coast of Egypt. M. Sc. Thesis, Fac Sci, Alexandria University

Barakat MG, Imam M (1976) Preliminary note on the occurrence of old indurated sand dunes in the district of Gamasa, northeastern Nile delta. In: Proceedings on seminar on Nile delta sedimentology, UNDP/ASRT/UNESCO, pp 33–39

Batisha AF (2012) Adaptation of sea level rise in Nile Delta due to climate change. Earth Sci Climate Change 3:p5

Bigio A (2009) Adapting to climate change and preparing for natural disasters in the coastal cities of North Africa. Urban Research Symposium. Cities and climate change: Responding to an urgent agenda, pp 28–30

Blodget H, Taylor PT, Roak JH (1991) Shoreline changes along the Rosetta-Nile Promontory: monitoring with satellite observations. Marine Geol 99:67–77

Butzer KW (1960) On the Pleistocene shorelines of Arabs' Gulf. Egypt J Geol 68:626–637

Cai F, Su X, Liu J, Li B, Lei G (2009) Coastal erosion in China under the condition of global climate change and measures for its prevention. Prog Nat Sci 19(4):415–426

CAPMAS (2014) Central agency for public mobilization and statistics

Chang K-t (2014) Introduction to Geographic Information Systems, 7th edn. McGraw Hill Higher Education, Boston, 425 pp

Chavez PS (1996) Image-based atmospheric correction revised and improved. Photogramm Eng Remote Sens 62:1025–1036

Cho Y-S et al (2001) A concept of beach protection with submerged breakwaters. J Coastal Res 34:671–678

Chust G, Caballero A, Marcos M, Liria P, Hernandez C, Borja A (2010) Regional scenarios of sea level rise and impacts on Basque (Bay of Biscay) coastal habitats, throughout the 21st century. Estuar Coast Shelf Sci 87:113–124

Coastal Research Institute (1993) Short-term studies and proposed measures for improving nearshore coastal zone conditions along Marakia tourist village. Technical report, Alexandria, Egypt, 31 pp

Coastal Research Institute (2000) Report on hydrographic profile survey along the northwestern coast of Egypt, West Alexandria to Sallum. Technical Report, Alexandria, 30 pp

Colwell RN (1983) Manual of Remote Sensing, 2nd edn, American Soc. Photogrammetry, Washington, DC, 2440 pp

Darwish K, Smith S (2021) A Comparison of Landsat-8 OLI, Sentinel-2 MSI and PlanetScope Satellite Imagery for Assessing Coastline Change in El-Alamein, Egypt. Eng Proc 10(1):23. https://doi.org/10.3390/ecsa-8-11258

Darwish K, Smith SE (2023) Landsat-based assessment of morphological changes along the Sinai Mediterranean coast between 1990 and 2020. Remote Sensing 15(5):1392. https://doi.org/10.3390/rs15051392

Darwish K, Smith S, Torab M, Monsef H, Hussein O (2017) Geomorphological changes along the Nile Delta Coastline between 1945 and 2015 detected using satellite remote sensing and GIS. J Coastal Res 33(4):786–794

Dasgupta S, Laplante B, Meisner C, Wheeler D, Yan J (2009) The impact of sea level rise on developing countries: a comparative analysis Clim. Change 93:379–388

Davidson MA, Lewis RP, Turner I (2010) Forecasting seasonal to multiyear shoreline change. Coast Eng 57(6):620–629

Dawod G, Mohamed H, Haggag G (2019) Relative and absolute sea level rise based on recent heterogeneous geospatial data: a case study in the Nile delta Egypt. Sci Eng Res 6:55–64

Deabes EA (2003) The study of sea level changes and currents at Rosetta and Damietta outlets and Abu Quir Bay. M. Sc. Thesis, Faculty of Science, Alexandria University, 242 pp

Dean RG, Chen R, Browder AE (1997) Full scale monitoring study of a submerged breakwater, Palm beach, Florida. J Coastal Eng 29:291–315

Delft hydraulics (2000) Technical report on El-Alamein-Marina resort erosion study. Netherlands, 61 pp

Dewidar K, Frihy OE (2010) Automated techniques for quantification of beach change rates using Landsat series along the northeastern Nile Delta, Egypt. J Oceanogr Marine Res 1(2):28–39. https://www.longdom.org/archive/ocn-volume-1-issue-2-year-2010.html

Dewidar K, Bayoumi S (2021) Forecasting shoreline changes along the Egyptian Nile Delta coast using Landsat image series and Geographic Information System. Environ Monit Assess 193:429

Dewidar K, Frihy O (2007) Preand postbeach response to engineering hard structures using Landsat time-series at the northwestern part of the Nile delta. Egypt J Coast Conserv 11(2):133–142. https://doi.org/10.1007/s11852-008-0013-z

Diab MS, Dahab KA, El-Fakharany MA (1997) Impact of paleohydrogeological conditions on the groundwater quality in the northern part of the Nile Delta, Egypt. Egyp J Geology 41(2B):779–796

Djouder F, Boutiba M (2017) Vulnerability assessment of coastal areas to sea level rise from the physical and socioeconomic parameters: case of the Gulf Coast of Bejaia, Algeria. Arab J Geosci 10:299. https://doi.org/10.1007/s12517-017-3062-5

Eckert JW (1983) Design of toe protection for coastal structures. In: Proceedings of the coastal structure 83. American Society of Civil Engineers, Alexandria, VA, pp 331–341

Eid HM, El-Marsafawy S, Ouda S (2006) Assessing the economic impacts of climate change on agriculture in Egypt: a Ricardian approach. CEEPA Discussion Paper No. 16

El Banna MM (2004) Nature and human impact on Nile delta coastal sand dunes. Egypt Environ Geol 45:381–390

El Banna M (2007) Erosion and accretion rates and their associated sediment characters along Ras El Bar coast, northeast Nile Delta, Egypt. Environ Geol 52:41–49
El Banna MM, Frihy OE (2009) Natural and anthropogenic influences in the northeastern coast of the Nile delta, Egypt. Environ Geol 57:1593–1602. https://doi.org/10.1007/s00254-008-1434-6
El Banna MM, Mahmoud M (2008) Vulnerability and fate of a coastal sand dune complex, Rosetta-Idku, northwestern Nile Delta, Egypt. Environ Geol 54(6):1291–1299
El Fawal FM, Shendi EH (1991) Sedimentology and groundwater of the Quaternary sandy layer north of Wadi El Tumilat, Ismailia, Egypt. Ann Geol Surv Egypt v(xvii):305–314
El Fishawi NM, Badr AA (1989) Volumetric changes of nearshore sediments between Rosetta and Burullus, Egypt. INQUA Comm Q Shoreline Newsl 11:39–42
El Nahry AH, Doluschitz R (2010) Climate change and its impacts on the coastal zone of the Nile Delta. Environ Earth Sci 59:1497–1506
El Raey M (2010) Impact of sea level rise on the Arab Region. Transport 1–89
El Raey M, Nasr S, Frihy OE, Desouk S, Dowidar K (1995) Potential impacts of accelerated sea level rise on Alexandria Governorate, Egypt. J Coastal Res 51:190–204
El Raey M, Dewidar K, El Hattab M (1999) Adaptation to the impact of sea level rise in Egypt. Clim Res J 12:117–128
El Sayed MK (1991) Implications of climatic changes for coastal areas along the Nile, Delta. Environ Prof USA 13:59–65
El Sayed MK (1988) Sea level rise in Alexandria during the late Holocene: archaeological evidence. Rapp Comm Int Mer Medit, p 31
El Sayed MA, Mahmoud S (2007) Groins System for Shoreline Stabilization on the East Side of the Rosetta Promontory, Nile Delta Coast. J Coast Res 23:380–387
El Wany MH, Khafagy AA, Inman DL, Fanos AM (1988) Analysis of waves from arrays at Abu Quir and Ras El Bar, Egypt. Adv Underw Technol Ocean Sci Offshore Eng 16:89–97
El-Ashmawy A (2012) Sarekate Mashroaa (in Arabic), 1st ed. Eldar El-Masria
El-Asmar HM (2002) Holocene paleoshorelines along the Manzala Lagoon, northeast of the Nile Delta, Egypt. N J Geol Paläonto, Stuttgart 6:337–361
El-Asmar HM, Hereher ME (2011) Change detection of the coastal zone east of the Nile Delta using remote sensing. Environ Earth Sci 62(4):769–777. https://doi.org/10.1007/s12665-010-0564-9. Accessed 25 Aug 2021
El-Asmar HM, Taha MMN, El-Kafrawy S (2014) Monitoring coastal changes along Damietta promontory and the barrier beach toward Port Said east of the Nile Delta. Egypt J Coast Res 30(5):993–1005. https://doi.org/10.2112/JCOASTRES-D-12-00112.1
El-Asmar HM, Taha MM, El-Kafrawy SB, El-Sorogy AS (2015) Control of Late Holocene geo-processes on the sustainable development plans of the Tineh Plain, NW Sinai coast, Egypt. J Coast Conserv 19:141–156. https://doi.org/10.1007/s11852-015-0377-9
El-Banna MM (2006) Responses of Ras El-Bar seafloor characteristics to the protective engineering structures, Nile Delta, Egypt. Environ Geol 49(5):645–652
El-Banna MM, Hasaneen NNA (1999) Erosion and accretion processes in response of protection structures on Ras El-Bar Resort, Damietta Governorate, Egypt. J Environ Sci Univ Mansoura Egypt 17:69–88
El-Banna MM, Hasaneen NNA (2002) Current pattern and bathymetry of Ras El-Bar Resort behind the breakwater system. J Environ Sci Univ Mansoura Egypt 23:119–132
El-Masry, EA (2017) Adaptation to the impacts of climate change on El Hammam—El Alamein area, Mediterranean coast of Egypt, within the framework of integrated coastal zone management (p. 135). PhD Thesis, Alexandria University, Egypt
El-Masry EA, El-Sayed MK, Awad MA, El-Sammak AA, El Sabarouti MA (2021) Vulnerability of tourism to climate change on the Mediterranean coastal area of El Hammam–El Alamein, Egypt. Environ Dev Sustain 24:1145–2116
El-Nahry AH (2001) An approach for sustainable land use studies of some areas in Northwest Nile Delta, Egypt. PhD thesis, Faculty of Agriculture, Soil Department, Cairo University, Egypt

El-Nahry AH, Doluschitz R (2010) Climate change and its impacts on the coastal zone of the Nile Delta, Egypt. Environ Earth Sci 59:1497–1506. https://doi.org/10.1007/s12665-009-0135-0

El-Raey M, Nasr S, Frihy O, Desouki S, Dewidar K (1995) Potential impacts of accelerated sea-level rise on Alexandria Governorate, Egypt. J Coast Res 14:190–204

El-Raey M, Sharaf El-Din SH, Khafagy AA, Abo Zed AI (1999) Remote sensing of beach erosion/accretion patterns along Damietta-Port Said shoreline, Egypt. Int J Remote Sens 20:1087–1106

El-Shaer MH, Eid HM, Rosenzweig C, Iglesias A, Hellil D (1996) Agricultural adaptation to climate change in Egypt. Meteorology and integrated development conference, March, Egypt, pp 24–25

Eldeberky Y, Hünicke B (2015) Vulnerability of the Nile delta to recent and future climate change. In: E-Proceedings of the 36th IAHR World Congress 28 June–3 July, 2015 Delft—The Hague, the Netherlands

Elsayed MA, Younan NA, Fanos AM, Baghdady KH (2005) Accretion and erosion patterns along Rosetta promontory, Nile Delta coast. J Coast Res 21:412–420

Elsharkawy H, Rashed H, Rached I (2009) Climate change: the impacts of sea level rise on Egypt. In: The impacts of SLR on Egypt, 45th ISOCARP Congress 2009, Porto, Portugal

ESRI (2007) Arc Map version 9.2. user manual. ESRI, 380 New York Street, Redlands, CA, 92373-8100, USA

Fairbanks RO (1989) A 17000 year glacioeustatic sea-level record: influence of glacial melting rates on the Younger Dryas event and deep ocean circulation. Nature 342:637–642

Fanos AM (1986) Statistical analysis of longshore current data along the Nile Delta coast. J Water Sci 1:45–55

Fanos AM (1996) Wave climate along the Mediterranean Egyptian Coast. Technical Progress Report No. W1. Coastal Research Institute, Alexandria, May 1996

Fanos AM, Frihy OE, Khafagy AA, Komar PD (1991) Processes of shoreline change along the Nile delta coast of Egypt. Proceedings of the Coastal Sediments '91 Conference 2:1547–1557

Fanos AM, Naffaa MG, Fouad EE, Omar W (1995) Seasonally and yearly wave regime and climate of the Mediterranean coast of Egypt. COPEDEC IV, Rio de Janeiro, Brazil

FAO (Food and Agriculture Organization) (2015) GIEWS Country Brief on Egypt. Available online at: http://www.fao.org/giews/countrybrief/country.jsp?code=EGY. Accessed 31 Dec 2017

Fourtau R (1893) La région de Maryut; étude géologique. Bull Inst Egypt Sér 3:1–141

Frihy OE (1992) Sea level rise and shoreline retreat of the Nile delta promontories, Egypt. Natural Hazards Netherlands 5:65–81

Frihy OE (2003) The Nile delta-Alexandria coast: vulnerability to sea-level rise, consequences and adaptation. Mitig Adapt Strateg Strateg Glob Change 8:115–138

Frihy OE (2009) Morphodynamic implications for shoreline management of the western Mediterranean sector of Egypt. Environ Geol 58:1177–1189

Frihy O (2017) Evaluation of future land-use planning initiatives to shoreline stability of Egypt's northern Nile delta. Arab J Geosci. https://doi.org/10.1007/s12517-017-2893-4

Frihy OE, El-Sayed MK (2013) Vulnerability risk assessment and adaptation to climate change induced sea level rise along the Mediterranean coast of Egypt. Mitig Adapt Strateg Glob Change 18:1215–1237

Frihy OE, Khafagy AA (1991) Climatic and human induced changes in relation to shoreline migration trends in the Nile Delta promontories. CATENA 18:197–211

Frihy OE, Komar PD (1991) Patterns of beach-sand sorting and shoreline erosion on the Nile delta. J Sediment Petrol 61:544–550

Frihy OE, Komar PD (1993) Long-term shoreline changes and the concentration of heavy minerals in beach sands of the Nile delta, Egypt. Marine Geol J 115:253–261

Frihy OE, Lawrence D (2004) Evolution of the modern Nile delta promontories: development of accretional features during shoreline retreat. Environ Geol 46:914–931

Frihy OE, Stanley DJ (1988) Texture and coarse fraction composition of Nile delta deposits: facies analysis and stratigraphic correlation. J Afr Earth Sci 7:237–255

Frihy OE, El Fishawi M, El Askary MA (1988) Geomorphological features of the Nile delta coastal plain: a review. Acta Adriat 29:51–65

Frihy OE, Nasr SM, El Hattab MM, El Raey M (1994) Remote sensing of beach erosion along the Rosetta Promontory, northwestern Nile delta, Egypt. Int J Rem Sens 15:1649–1660

Frihy O, Dewidar K, Nasr S, El Raey M (1998a) Change detection of the northern Nile delta of Egypt: shoreline changes, Spit evolution, margin changes of Manzala lagoon and its islands. Int J Remote Sens 19:1901–1912

Frihy OE, Dewidar K, El Banna M (1998b) Natural and human impact on the northeastern Nile Delta coast of Egypt. J Coast Res 14:1109–1118

Frihy OE, Badr AM, Hassan MS (2002) Sedimentation processes at the navigation channel of the Damietta Harbor on the Northeastern Nile Delta coast of Egypt. J Coast Res 18:459–469

Frihy O, Dewidar K (2003) Patterns of erosion/sedimentation, heavy mineral Concentration and grain size to interpret boundaries of littoral sub-cells of the Nile delta. Marine Geology 199, 27–43

Frihy O, Debes A, El Sayed W (2003) Processes reshaping the Nile Delta promontories of Egypt: pre- and postprotection. Geomorphology 53(3–4):263–279

Frihy OE, El Banna MM, El Kolfat AI (2004a) Environmental impacts of Baltim and Ras El Bar shore-parallel breakwater systems on the Nile delta littoral zone, Egypt. Environ Geol 45:381–390

Frihy OE, Iskander MM, Moniem AE, Badr A (2004b) Effects of shoreline and bedrock irregularities on the morphodynamics of the Alexandria coast littoral cell, Egypt. Geo-Mar Lett 24:195–211

Frihy OE, Deabes EA, Gindy AAE (2010a) Wave climate and nearshore processes on the Mediterranean coast of Egypt. J Coastal Res 26(1):103–112. JSTOR

Frihy OE, Deabes EA, Shereet SM, Abdalla FM (2010b) Alexandria-Nile delta coast, Egypt: Update and future projection of relative sea-level rise. Environ Earth Sci 61:253–273. https://doi.org/10.1007/s12665-009-0340-x

Frihy OE, Deabes E, Moufaddal WM, El-Shahat A (2015) Recycling of coastal dredged sediments from the northern Nile delta, Egypt, for heavy minerals exploitation. Mar Georesour Geotechnol 33:408–418

Frihy O, Moufaddal W, Deabes E, Helmy D (2016) Economic evaluation of using marine dredged material for erosion control along the northeast coast of the Nile Delta. Arab J Geosci. https://doi.org/10.1007/s12517-016-2660-y

Frihy O, Mohamed S, Abdalla D, El-Hattab M (2021) Assessment of natural coastal hazards at Alexandria/Nile Delta interface, Egypt. Environ Earth Sci 80:1–15

Gad MA, Saad A, El-Fiky A, Khaled M (2013) Hydrodynamic modeling of sedimentation in the navigation channel of Damietta Harbor in Egypt. Coastal Eng J 55(2):1350007-1–1350007-31. https://doi.org/10.1142/S0578563413500071

Geirnaert W, Laeven MP (1992) Composition and history of groundwater in the western Nile Delta. J Hydrol 138:169–189

Geriesh MH, Balke KD, El-Rayes AE, Mansour BM (2015) Implications of climate change on the groundwater flow regime and geochemistry of the Nile Delta, Egypt. J Coast Conserv 19:589–608

Geriesh MH (2000) Paleohydrogeological regime of groundwater flow in the eastern Nile Delta Region, Egypt. In: Proceedings of the 4th international conference on water supply and water quality, Krakow, Poland, pp 229–241

Ghoneim E, Jehan M, Douglas G, Joanne H, Mostafa A (2015) Nile Delta exhibited a spatial reversal in the rates of shoreline retreat on the Rosetta promontory comparing pre- and postbeach protection. Geomorphology 228:1–14

Guth PL (2010) Geomorphometric comparison of ASTER GDEM and SRTM. ASPRS, ASPRS/CaGIS, Orlando, Florida

Hallermeier RJ (1981) A profile zonation for seasonal sand beaches from wave climate. Coastal Eng 4:253–277

Hamed AA, El Gindy AA (1988) Storm surge generation by winter cyclone at Alexandria, Egypt. J Int Hydrogr Rev Monaco 36:129–139

Hansen J, Sato M, Hearty P, Ruedy R et al (2016) Ice melt, sea level rise and superstorms: evidence from paleoclimate data, climate modeling, and modern observations that 2 degree C global warming could be dangerous. Atmos Chem Phys 16:3761–3812

Hasan E, Khan SI, Hong Y (2015) Investigation of potential sea level rise impact on the Nile Delta, Egypt using digital elevation models. Environ Monitor Assess 187(10):649. https://doi.org/10.1007/s10661-015-4868-9

Hassaan MA (2013) GIS-based risk assessment for the Nile Delta coastal zone under different sea level rise scenarios case study: Kafr EL Sheikh Governorate, Egypt. J Coast Conserv 17(4):743–754. https://doi.org/10.1007/s11852-013-0273-0

Hassaan MA, Abdrabo MA (2013) Vulnerability of the Nile Delta coastal areas to inundation by sea level rise. Environ Monitor Assess 185(8):6607–6616. https://doi.org/10.1007/s10661-012-3050-x

Herbich JB (ed) (1991) Handbook of coastal and ocean engineering. Gulf Publishing, Houston, Texas, 1340 pp

Hereher M (2010) Vulnerability of the Nile Delta to sea level rise: an assessment using remote sensing. Geomat Nat Hazards Risk 1(4):315–321

Hereher ME (2011) Mapping coastal erosion at the Nile Delta western promontory using Landsat imagery. Environ Earth Sci 64:1117–1125

Hereher ME (2014) The Lake Manzala of Egypt: an ambiguous future. Environ Earth Sci 72(6):1801–1809

Hilmy ME (1951) Beach sands of the Mediterranean coast of Egypt. Sediment Petrol 21:120

http://www.clac.edu.eg/Arabic/El%20watany/Egypt_INC.pdf. Accessed 15 Feb 2018

Ibrahim MM (1963) The last subsidence movement of land on the Mediterranean coast, unpublished report, Mineral Wealth and Ground Water. Ministry of Scientific Research, Egypt, p 12

IDSC (Egypt's Cabinet Information and Decision Support Centre). (2011). Statistical guide for Matrouh Governorate, June

IDSC/UNDP (2011) Egypt's national strategy for adaptation to climate change and disaster risk reduction. The Egyptian Cabinet, Information and Decision Support Centre, Cairo, p 166

IH Cantabria/Environics (2017) Integrated coastal zone management in the Northern Coast of Egypt, A Scoping Study. Draft of the final report. Adaptation to climate change in the Nile Delta through integrated coastal zone management project, pp 1–62

Inman DL, Jenkins SA (1984) The Nile littoral cell and man's impact on the coastal zone of the southeastern Mediterranean. Scripps Institution of Oceanography, Reference, References Series 84–31, University of California, La Jolla, 43 pp

Inman DL, Hany SW, Khafagy AA, Abraham G (1992) Nile Delta profiles and migrating sand blankets. In: Proceedings 23rd conference on coastal engineering. American Society of Civil Engineering, pp 3273–3284

IPCC (1990) Climate change: the IPCC scientific assessment: sea level rise. In: Jenkins GJ, Ephraums JJ (eds) Houghton JT. Cambridge University Press, New York, pp 257–281

IPCC (2001) Climate change 2001: impacts, adaptation, and vulnerability—contribution of working group II to the IPCC Third Assessment Report, Cambridge University Press, 572 pp

IPCC (2007) Climate Change 2007: Impacts, adaptation and vulnerability. Contribution of working group II to the third assessment report of the intergovernmental panel on climate change. In: Parry ML, Canziani OF, Palutikof JP, van der Linden PJ, Hanson CE (eds). Cambridge University Press, Cambridge, UK, pp 1000

IPCC CZMS (1992) Global climate change and the rising challenge of the sea. Workshop, Isla de Margarita, Venezuela, 9–13 March 1992

Iskander MM (2013) Wave climate and coastal structures in the Nile Delta coast of Egypt. Emirates J Eng Res 18(1):43–57

Iskander MM, Frihy EO, El-Anssary AE, Abd El-Mooty MM, Nagy HM (2007) Beach Impacts of shore-parallel breakwaters backing offshore submerged ridges, western Mediterranean coast of Egypt. J Environ Manage 85:1109–1119

Iskander MM (2010) Environmental friendly methods for the Egyptian coastal protection. In: Coastal zone management of river deltas and low land coastlines, pp 6–10

ITT (2008) ITT corporation ENVI 4.5 software,1133 Westchester Avenue, White Plains, NY 10604, USA

Jahn R, Blume HP, Asio VB, Spaargaren O, Schad P (2006) Guidelines for soil description. FAO, Rome

Khafagy AA, Manohar M (1979) Coastal protection of the Nile Delta. Nat Resour 15:7–13

Klemas VV (2015) Coastal and environmental remote sensing from unmanned aerial vehicles: an overview. J Coastal Res 31(5):1260–1267

Kotb M, Ali R, El Semary M (2017) Use of remote sensing and GIS for land degradation assessment of Qarun lake coastal area, El-Fayoum, Egypt. In: Water and Land Security in Dry lands. https://doi.org/10.1007/978-3-319-54021-4_22

Krumbein WC, James WR (1965) A lognormal size distribution model for estimating stability of beach fill material. Coastal Engineering Research Center, US Army Eng Waterw Exp Station, Vicksburg, MS, Rep TM-16

Li J (2020) Satellite remote sensing technologies. Springer

Lindell LT, Alexandersson ET, Norman JO (1991) Satellite mapping of oolitic ridges in Arabs Gulf, Egypt. J Geocarto Int Sweden 1:49–60

Link PM, Piontek F, Scheffran J, Schilling J (2012) On foes and flows: vulnerabilities, adaptive capacities and transboundary relations in the Nile River Basin in times of climate change. L'europe En Form 365:99–138

Louati M, Saïdi H, Zargouni F (2014) Shoreline change assessment using remote sensing and GIS techniques: a case study of the Medjerda delta coast, Tunisia. Arab J Geosci. https://doi.org/10.1007/s12517-014-1472-1

Lowe JA, Gregory JM (2005) The effects of climate change on storm surges around the United Kingdom. Philos Trans R Soc A: Math Phys Eng Sci 363(1831):1313–1328

Lunetta RS, Elvidge CD (1998) Remote sensing change detection: environmental monitoring methods and applications. Ann Arbor Press, Michigan

Manohar M (1981) Coastal processes on the Nile Delta coast. J Shore Beach 49:8–15

Manohar M (1976) Beach profiles. In: Proceedings of the UNESCO seminar on Nile Delta sedimentology, Academy of Scientific Research and Technology, Alexandria, pp 95–99

MASR (1978) El-Behaira Governorate Soil Fertility, Ministry of Agriculture and Soil Reclamation, Final report

Masria A, Iskander M, Negm A (2015) Coastal protection measures, case study (Mediterranean zone, Egypt). J Coast Conserv 19(3):281–294. https://doi.org/10.1007/s11852-015-0389-5

Masria A, El-Adawy AA, Sarhan T (2020) A holistic evaluation of human-induced LULCC and shoreline dynamics of El-Burullus Lagoon through remote sensing techniques. Innov Infrastruct Solut 5(3):1–10. https://doi.org/10.1007/s41062-020-00331-w

Mass JF (1999) Monitoring land-cover changes: A comparison of change detection techniques. Int J Remote Sens 20:139–152

Mieczkowski Z (1985) The tourism climate index: A method for evaluating world climates for tourism. Can Geogr 29:220–233

Mikhailova MV (2001) Hydrological regime of Nile Delta and dynamics of its coastline. Water Resour 28(5):477–490

Milliman JD, Broadus JM, Gable F (1989) Environmental and economic implications of rising sea-level and subsidence deltas: The Nile and Bengal examples. Ambio 18:340–345

Ministry of State for Environmental Affairs (MSEA) (2001) National Environmental Action Plan (NEAP) 2002–2017. MSEA, Cairo

Nafaa MG, Fanos AM, Khafagy AA (1991) Characteristics of waves off the Mediterranean coast of Egypt. J Coast Res 7:665–676

Nicholls RJ, Hoozemans FM, Marchand M (1999) Increasing flood risk and wetland losses due to global sea-level rise: regional and global analyses. J Global Environmental Change 9:S69–S87

O'Keefee PD (1978) Sediment budgeting, beach conservation, Beach Protection Authority, Queens Land, Issue No. 30

Oswald MR, Treat C (2013) Identifying sea level rise vulnerability using GIS: Development of a transit inundation modeling method. Int J Geoinformatics 9:1–10

Perlin M, Dean RG (1983) A numerical model to simulate sediment transport in the vicinity of coastal structures. US Army Corps of Engineers, CERC, Fort Belvoir, Misc Rep 83–10

Pilarczyk KW (2003) Design of low-crested (submerged) structures–an overview. In: 6th International Conference on Coastal and Port Engineering in Developing Countries, Colombo, Sri Lanka

Pirazzoli PA (1991) World Atlas of Holocene sea-level changes. Elsevier Oceanography Series 58, Amsterdam, 300 pp

Poulter B, Halpin PN (2008) Raster modeling of coastal flooding from sea-level rise. Int J Geogr Inf Sci 22(2):167–182

Racha R (2015) Demand and supply challenges of food security in Egypt, Review No. 2, Economic Literature Review. www.eces.org.e.g./MediaFiles/Uploaded_Files/4fcd3c72.pdf

Randazzo D, Stanley D, Geronimo I, Amore C (1998) Human-induced sedimentological changes in Manzala Lagoon, Nile Delta. Egypt. Environ Geol 36:235–258

Raymond C, Conway-Cranos L, Morgan H, Faghin N, Spilsbury Pucci D, Krienitz J, Miller I, Grossman E, Mauger G (2018) Sea level rise considerations for near-shore restoration projects in Puget Sound. A report prepared for the Washington Coastal Resilience Project. USA: University of Washington. https://cig.uw.edu/wp-content/uploads/sites/2/2018/11/Raymondetal_SLR_Restoration_2018_Compressed.pdf. Accessed Oct 2020

Ritchie JC, Cooper CM, Schiebe FR (1990) The relationship of MSS and TM digital data with suspended sediments, chlorophyll, and temperature in Moon Lake, Mississippi. Remote Sens Environ 33:137–148

Rosen DS, Vajda M (1982) Detached breakwaters. Proc 18th Int Conf Coastal Eng 1930–1949

Said R (1993) The evolution of the Nile Delta. Geoscientific research in Northeast Africa, Thorwehe and schandelmeir (eds) Balkema R, pp 615–619

Said R, Philip G, Shukri NM (1956) Post Tyrrhenian climatic fluctuations in northern Egypt. Quaternaria 3:167–172

Sestini G (1976) Geomorphology of the Nile Delta. Seminar on Nile Delta sedimentology, CoRI/UNESCO/UNDP, Alexandria, pp 12–24

Sestini G (1989) Nile Delta: A review of depositional environments and geological history. Geol Soc Spec Pub 41(1):99–127. https://doi.org/10.1144/GSL.SP.1989.041.01.09

Sestini G (1992) Implications of climatic changes for the Nile delta. In: Jeftic L, Millimam JD, Sestini G (eds) Climate change and the mediterranean. Edward Arnold, London, pp 535–557

Shukri NM (1950) The mineralogy of some Nile sediments. QJGS Lond 105:511–534

Shukri NM, Philip G, Said R (1956) The geology of the Mediterranean coast between Rosetta and Bardia. Part II. Pleistocene sediments: geomorphology and microfacies. Bull Inst Egypt 37:395–427

Song C, Woodcock CE, Seto KC, Lenney MP, Macomber SA (2001) Classification and change detection using Landsat TM data: when and how to correct atmospheric effects. Rem Sens Environ 75:230–244

Stanley DJ (1988a) Low sediment accumulation rates and erosion on the middle and outer Nile delta shelf off Egypt. Mar Geol 84(1):111–117. https://doi.org/10.1016/0025-3227(88)90129-6

Stanley DJ (1988b) Subsidence in the northeastern Nile delta: rapid rates, possible causes, and consequences. Science 240(4851):497–500

Stanley DJ (1990) Recent subsidence and northeast tilting in the Nile Delta. Egypt Mar Geol 94:147–154

Stanley JD, Clemente PI (2017) Increased land subsidence and sea-level rise are submerging Egypt's Nile Delta coastal margin. GSA Today. https://doi.org/10.1130/GSATG312A.1

Stanley DJ, Goodfriend GA (1997) Recent subsidence of the northern Suez canal. Sci Corresp 388:335–336

Stanley JD, Jorstad TF, Goddio F (2006) Human impact on sediment mass movement and submergence of ancient sites in the two harbours of Alexandria, Egypt. Norwegian J Geol 86:337–350

Stanley DJ, Warne AG (1993) Nile delta: recent geological evolution and human impacts. Sci J 260:628–634

Stanley DJ, Warne AG (1998) Nile delta in its destruction phase. J Coastal Res 14:794–825

Steinberg SL, Steinberg SJ (2015) GI research methods: incorporating spatial perspectives. Esri Press, Redlands, CA. ISBN 978-1-58948-378-1

Strzepek KM, Yates DN, El-Quosy DED (1995) Vulnerability assessment of water resources in Egypt to climatic change in the Nile Basin. Clim Res 6(2):89–95

Syvitski JPM (2008) Deltas at Risk. Sustain Sci 3:23–32

Szmytkiewicz M, Biegowski J, Kaczmarek LM, Okroj T, Ostrowski R, Pruszak Z, Rózy'nsky G, Skaja M (2000) Coastline changes nearby harbor structures: comparative analysis of one-line models versus field data. Coast Eng 40(2):119–139

Taha MM, El-Asmar HM (2018) Geo-archeoheritage sites are at risk, the Manzala Lagoon, NE Nile Delta Coast, Egypt. Geoheritage Online First 2018:1–17. https://doi.org/10.1007/s12371-018-0297-9

Thieler ER, Himmelstoss EA, Zichichi JL, Miller TL (2005) Digital Shoreline Analysis System (DSAS) version 3.0; an Arc GIS extension for calculating shoreline change. U.S. Geological Survey Open-File Report, 2005-1304

Tian B (2017) GIS technology applications in environmental and earth sciences, 1st edn. CRC Press, Boca Raton

Torab M, Azab M (2006) A modern shoreline changes along the Nile Delta coast as an impact of construction of the Aswan High Dam, World Congress of Soil Science, Philadelphia, Pennsylvania, USA, July 9–15, 2006

Turner RK (2000) Integrating natural and socioeconomic science in coastal management. J Mar Syst 25(3–4):447–460

UNDP (1978) Coastal protection studies. Final technical report, Paris, p 155

UNDP (2009) Adaptation to Climate Change in the Nile Delta through Integrated Coastal Zone Management. United Nations Development Programme, Project Document 3748, New York, 124 pp

UNDP-RBAS (2012) UNDP Regional Bureau of Arab States, Impacts of sea level rise on Arab region. http://www.arabclimateinitiative.org/Countries/egypt/ElRaey_Impact_of_Sea_Level_Rise_on_the_Arab_Region.pdf. Accessed on 10 May 2012

UNESCO/UNDP (1978) Coastal protection studies. Final Technical Report No. 1. UNESCO/UNDP, Paris, 155 pp

US Army Crops of Engineers (1984) Shore protection Manual, vol 1, Chapter 7. Coastal Engineering Research Center, Washington, DC

Van Rijn LC (2004) Estuarine and coastal sedimentation problems. In: Proceedings of the Ninth International Symposium on River Sedimentation. pp 105–120

Wang Y (2010) Remote sensing of coastal environments: an overview. In: Wang J (ed), Remote sensing of coastal environments. CRC, Boca Raton, Florida, 423 pp

Wang Y, Moskovits DK (2001) Tracking fragmentation of natural communities and changes in land cover: Applications of Landsat data for conservation in an urban landscape (Chicago Wilderness). Conserv Biol 15(4):835–843

Wang Y, Brian R, Mitchell BM, Marzilli JN, Bonygne G, Zhou Y, Shriver G (2009) Remote sensing of land-cover change and landscape context of the National Parks: a case study of the Northeast Temperate Network. Rem Sens Environ 113:1453–1461

Wiggel RI (1964) Oceanographic engineering: Prentice Hall, Englewood Cliffs, 532 pp

Zaghloul ZM, Taha HH, Hegab OA and El Fawal FM (1979) The Plio-Pliostocene Nile Delta subenvironments; stratigraphic section and genetic class. Geol Soc Vol. 1x, Cairo

Chapter 2
Physical Characteristics of the Nile Delta's Coastal Plain

Abstract Physical characteristics are considered essential driving factors for understanding coastal morphodynamics, processes, and dynamics and assessing potential disasters, hazards, and risks. This chapter aims to assess the physical characteristics of the NDC; three different physical characteristics were analyzed and assessed: first, the geological setting, which includes the spatial distribution of the geological formations and surface deposits, and the undersurface geological lineaments. The morphological characteristics of NDC are the second physical factor, including topographical analysis, slope zones, and aspects. The climatological characteristics are the last factors that affect the vulnerability of the NDC; they include climatological elements such as temperature, wind, storms, and rainfall.

Keywords Nile Delta Geology · Topography of Nile Delta · Nile Delta Geomorphology

2.1 General

Global coastal hazards and disasters threaten global deltaic zones due to their low topography, high risk of flooding, extensive land erosion, and high susceptibility to climatic changes (Milliman et al. 1989). Earth surface processes are responsible for producing distinct landforms and evolution, and the NDC is a highly dynamic landscape (Zuidam et al. 1998; Abou El-Magd and Hermas 2010). Coastal, Aeolian, human, and hydrodynamic processes are all part of these processes (Allen 1997). Coastal and hydrodynamic processes create several coastal dynamic landforms, such as sand beaches, coastal plains, salt marshes, wetlands, and water bodies. Natural coastal processes are causing catastrophic erosion along the Nile Delta's coastal zone. It is predicted that climate change has increased the rate of erosion along Egypt's Nile Delta coast. Moreover, rising sea levels could increase the risk of coastal inundation, erosion, and saltwater intrusion into freshwater aquifers and rivers in low-lying deltas in all places (El-Nahry and Doluschitz 2010). Understanding the geological setting of the Nile Delta is very important to assess land subsidence in the coastal zone, as well

K. S. Darwish, *Hazard Modeling and Assessment of the Nile Delta Coast*,
https://doi.org/10.1007/978-3-031-44324-4_2

as seawater intrusion in underground water aquifers. In this chapter, the significant physical characteristics that affect the vulnerability of NDC are analyzed, including geology, morphology, and climatology.

2.2 Geological Characteristics

The Nile Delta is one of the oldest known deltas in the world; it occupies an area of 22,000 km^2 and is located on the northeastern margins of the African plate on the southeastern side of the Mediterranean Sea (Farouk et al. 2014). The geological knowledge of the Nile Delta is still limited because the area does not contain ancient rock outcrops, and it is covered by Holocene soils (Barakat 2010). The formation of the Nile Delta began in the late Pliocene, but the main evolution occurred in the Pleistocene with association with the large volumes of coarse-grained coastal sands and accumulation of the Nile Cone (Sestini 1989).

The Nile Delta cone formed during the period from Neogene to Pleistocene from continental sediments. Moving down to the slope of the continental shelf due to gravitational forces on the coast, the rates of accumulation of cone sediments increased during the late Miocene until the late Quaternary when the river discharged its load directly on the upper slope when the sea level decreased (Bartolini et al. 1975). Large sedimentary forms resulting from the accumulation of the Nile River load for more than five million years. Irregular topographic forms form the surface of the cone affected by isostatic subsidence, faults, and local tectonic forms (Stanley and Warne 1998). It makes sense that the Nile Delta began to form during a time when sea levels were decreasing. That is because rivers deposited more material than the seas could take away, so the delta began to build up. That period was between 8000 and 6500 years BP, and it was the same time other deltas around the world were forming as well (Stanley and Warne 1994). The recent formation of the Nile Delta started in the Messinian Period, when the river made its way north from the African lakes plateau and traveled through Sudan, to Central Sudan, and Ethiopia before reaching Upper and Lower Egypt (Sestini 1989). The sea levels were lower during that time, so the Nile River's discharge spread over a wide area of land (Stanley and Warne 1998). This contributed to the creation of an active flood plain with river channels branching in many directions approximately 15–8 thousand years ago, and sea levels started to rise quickly, which caused much sediment to move around. Eight thousand years ago, sea levels slowed down, which led to the advance of the modern-day delta (Said 1990, 1993).

The Delta grew steadily with sediment deposition of 1–7 mm/year resulting in up to 60 m of sedimentary rocks in the Holocene Delta (Stanley and Warne 1993). The modern-day delta is located south of the Nile Cone, which was formed by more than 5 million years of discharge from both the Nile and paleo-Nile drains (Sestini 1989). According to Rizzini et al. (1978), the Neogene-Quaternary section in the delta consists of three sedimentation cycles: Miocene, Plio-Pleistocene, and Holocene. It was further separated into two units called the Mit-Ghamr and Bilqas Formations.

2.2.1 *Formation and Evolution of the Nile Delta Coastline*

The geological evolution of the NDC consists of five consequence stages over the late Quaternary according to Stanley and Warne (1993), summarized as follows (Fig. 2.1):

- During the first stage, from 35,000 to 18,000 years ago, most of the area was an alluvial plain with seasonally active braided channels. The sea level was low, which caused flood-plain mud to accumulate in ephemeral, seasonally dry depressions. In the west of the region, there were deposits of sand and mud from a generally arid climate that built up between and around carbonate ridges.
- In the second stage, it appears that the sea level increased quickly between 15,000 and 8000 years ago. This caused the shoreline to move inland, and the sand from the alluvial plain was reworked. It appears that the northern delta region may have been tilting toward the northeast around that time. However, we do not know where the Nile channels were during this period.
- In the third stage, the evolution of the Nile Delta's coast was marked by the formation of a modern delta approximately 75,000 years ago. This was due to sea level deceleration and an influx of sediment, so reworking by waves and currents was limited. At that time, the sea level was 9–10 m lower than the present day, the river gradient was steeper, and the climate was more humid. The morphology of the delta and facies distribution were mainly controlled by the Sebennitic channel, which transported large volumes of sand to the coast. This sand created an extensive accreted beach ridge system with a cuspate shape at the headland of the

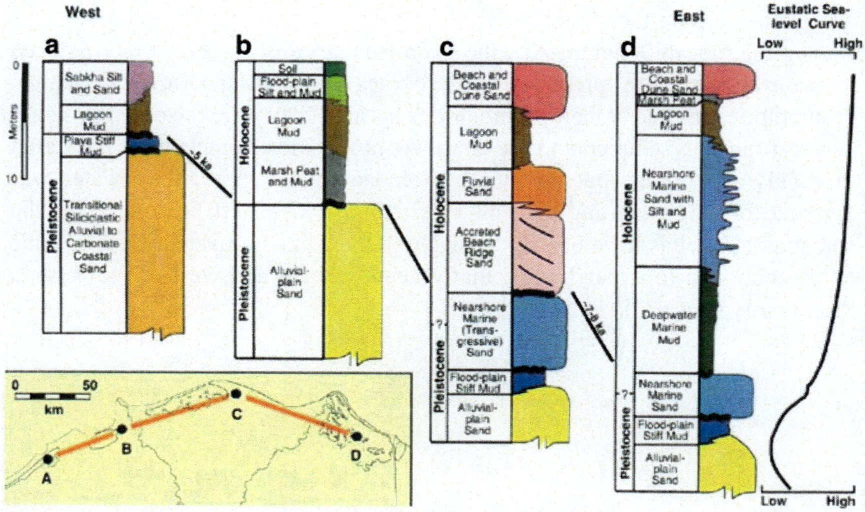

Fig. 2.1 Vertical Cross Sections for the Late Quaternary Stratigraphy in different locations along NDC, after (Stanley and Warne 1993)

delta. Sand ridges developed a barrier along the seaward borders of lagoons and marshes, and lobes began to form at major distributary mouths such as Mendesian and Pelusian in the Manzala lagoon area. The northeast tectonic tilt of this delta helped preserve these deposits, while distributaries westward, such as Sebennitic and Canopic, have not been recovered.

- Stage four started approximately 4000 years ago, and sea levels continued to rise, but more slowly, the slope of the delta plain decreased, and the climate became more arid. This period records the transition from a river-dominated, cusp-current-dominated, peaked delta to a wave-dominated, and arcuate delta. The northeastern sector continued to prograde, while the north-central coast began to decline. People who had settled in the delta as early as the Predynastic period established important population centers such as Buto and Menshat Abu Omar. Nonetheless, wetlands remained the primary ecosystem in the northern delta during the early to mid-pharaonic period.
- The fifth stage started approximately 2000 years ago. In this stage, the sea level rose to approximately 2 m below current levels, and ocean waves and currents shaped the shoreline so that the delta margin configuration began to resemble the modern shoreline. By this time, the delta had become more wave-dominated and had a gentle, arcing shape. However, at least five tributaries remained, most with small promontories, which continued to transport significant amounts of sediment to the coast during the annual floods. Extensive coastal dune fields developed in the north-central delta from eroded Sebennitic foothill sediments. At this time, humans significantly influenced Delta evolution: population increased in the Delta during the Hellenistic period, particularly in the Alexandria-Naukratis sector; the Damietta (Bucolic) and Rosetta (Bolbitine) canals were obtained by artificial excavations; and intensified wetland irrigation and drainage projects significantly altered the delta surface (Said 1981).
- During the first millennium AD, the main tributaries of the Nile were reduced to two, the artificially preserved distributaries Damietta and Rosetta. Distinct promontories formed at their promontories because they were the only two channels transporting sediments to the coast. As other side channels were converted to canals and drains that no longer extended to the shore, their water was diverted for irrigation, and the flow was reduced. The wetlands of the middle and southern deltas have been extensively drained and cultivated. Coastal dune fields continued to expand, particularly near Balteam and west of the Rosetta Promontory (Fig. 2.2).

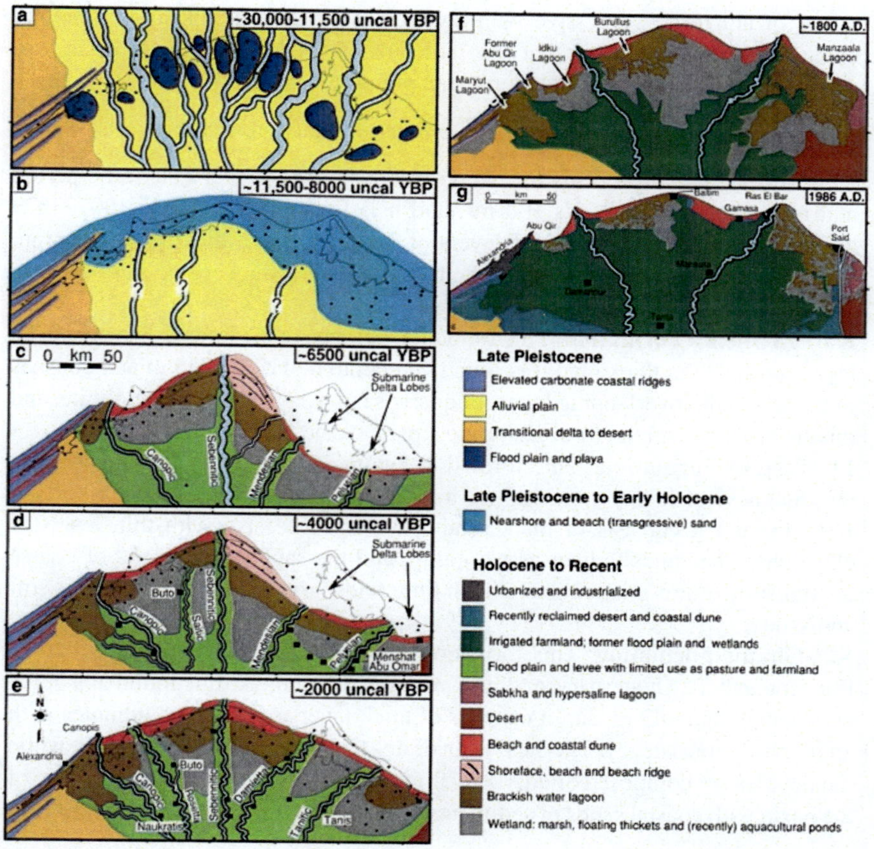

Fig. 2.2 Late Quaternary Paleogeography of NDC, after (Stanley and Warne 1993)

2.2.2 *Stratigraphy*

Rizzini et al. (1978) identified three different sedimentary cycles along the NDC: the Miocene cycle, the Paleocene-Pleistocene cycle, and the Holocene cycle.
The Miocene Cycle includes three geological formations:

- **Sidi Salim Formation**: The thickness of this formation is more than 700 m, it is located at a depth between 3592 and 4038 m and consists mainly of clay with a mixture of dolomite, marl, and stratified sandstone and mudstone, and the color ranges from gray to gray green.
- **Qawasim Formation**: It belongs to the depositional myosin cycle and is 700 m thick, and it is found at a depth of 2800–3733 m, consisting of successive irregular layers of sand, sandstone, and clay.
- **Rosetta Anhydrite Formation**: This formation belongs to the deposition myosin cycle. It is 50 m thick, located at a depth of 2678–2718 m, and consists of thick

layers of anhydrite successive with layers of clay. This composition does not contain fossils but reflects that the sediment of the Mediterranean basin was defective in the Messinian Mediterranean period.

The Paleocene-Pleistocene Cycle includes four geological formations:

- **Abu Madi Formation**: This formation belongs to the Lower Paleocene period and is. Its layer thickness is 300 ms, and it is found at a depth of 3007–3329 m. This composition consists of layers of very large and thick sand, with rare conglomerate matched with layers of clay that makes it thicker, and sand consists of quartz of various sizes poorly sorted.
- **Kafr El Sheikh Formation**: This formation belongs to the middle Paleocene. Its layers are 1500 m thick, and it is found at a depth of 1277–2735 m and consists of fine clay with overlapping layers of quartz sand. Clay consists of kaolinite, and this formation is spread throughout most parts of the Nile Delta.
- **El Wastani Formation**: This formation belongs to the Upper Paleocene era, the thickness of its layers is 300 m, it is located at a depth between 1009 and 1132 m, and it consists of thick sandy layers. Successive with thin levels of clay, which becomes thinner as we go up, and the sand is composed of coarse quartzose-medium-grained grains with little feldspar fractions, while clay is soft and sandy.
- **Mit-Ghamr Formation**: This formation belongs to the Pliocene period Upper Pliocene and the Quaternary, and its layers are 700 m thick. It is found at a depth between 20 and 483 m, and it consists of thick layers of sand and granules with an identical consistency. The sand is medium- to coarse-grained quartzose, while the pebbles are composed of quartzite, flint, and dolomite, reflecting the filling of the basin with coastal sands or sediments brought with the flood of the Nile River.

The Holocene Cycle, which includes:

- **Bilqas Formation**: This formation is 50 m thick, is found at a depth ranging between 0 and 25 m, and consists of successive medium-fine sands with clay rich in artifacts: Pelecypod, Gastropod, and Ostracode. The clay contains vegetable crumbs. The detectors of this formation appear throughout the delta mixed with the recent continental formations (Fig. 2.3).

2.2.3 The Surface Quaternary Deposits

Geologically, the Nile Delta also includes the continental shelf from 80 km west of Alexandria, approximately from the town of Al-Hamam to the village of Palouse in N Sinai, the continental slope, and the submarine Nile Cone (Sestini 1989). Analysis of the 1:500,000 scale geological maps published by the collaboration between Conoco Coral co. and the Egyptian Geological Survey in 1987 for Nile Delta coast; Sheets NH35NE Alexandria and NH36NW Cairo indicate that the coastal plain of the Nile Delta, covered by seven different geological surface deposits, belongs to

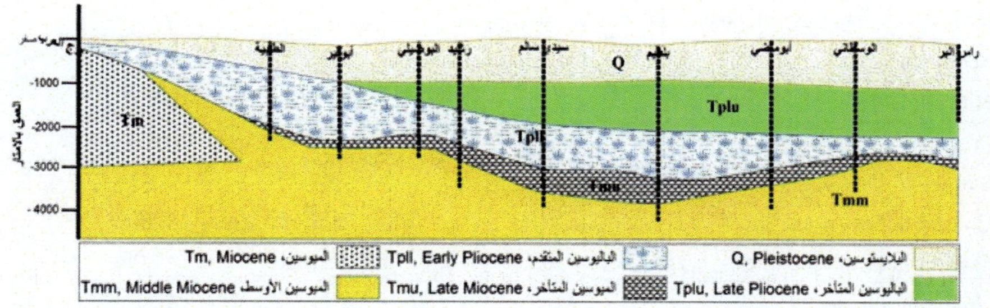

Fig. 2.3 Stratigraphic section of the study area from the west (Burj Al-Arab) to the east (Ras El Bar)

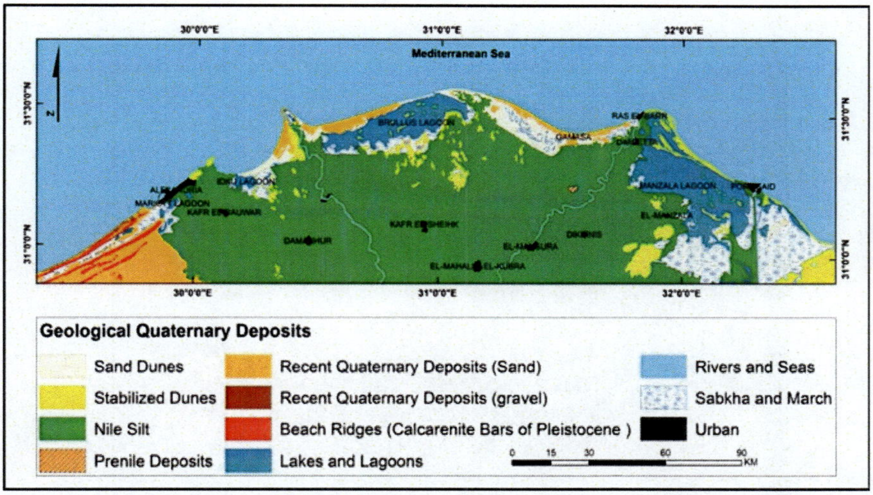

Fig. 2.4 Quaternary deposits Northern Nile Delta

the Quaternary, specifically the late Pleistocene and the Holocene. It was distributed along the coastal plain of the Nile Delta: Pleistocene calcarenite ingots (calcareous, marl, and oolitic beach ridges), Nile silts, Prenile deposits, undifferentiated Quaternary deposits (sands, gravels, and more recent coastal deposits), Sabkha and March deposits, sand dunes, and stabilized dunes, as shown in Fig. 2.4.

2.2.3.1 Nile Silt

It is considered one of the newest deposits in the study area, and it is alluvial sediment brought suspended by water. Nile River during the Holocene (Chen et al. 1992), and these sediments covered an area of 9972.06 km^2, or 72.98% of the total area of the

Table 2.1 Quaternary deposits in the study area

Sediment type	Geological sign	Area	Percent %
Nile silt	Qns	9972.06	72.98
Prenile	Qn2	6.83	0.05
Stabilized sand dunes	Qds	869.38	6.36
Coastal dunes	Qd	517.13	3.78
Sabkhas and salt marches	Qb	1510.16	11.05
Quaternary deposits	Q	671.06	4.91
Coastal ridges	Cr	44.16	0.33
Calcarenite bars of pleistocene age	Qc	73.69	0.54
Total area		13,664.47	100%

Fig. 2.5 Photograph showing an outcrop section of Nile silt

four sediments, as shown in Table 2.1, Fig. 2.5 which shows an outcrop of Nile silt deposits.

2.2.3.2 Prenile Deposits

They are thick sediments and easily recognizable units consisting mainly of sand, which are coarse and thick in the form of lumps, and it is likely that their source is from outside Egypt (Said, 1981, p. 44). These deposits appear in one location in the coastal plain of the Nile Delta, south of Gamasa, approximately 25 km, near the

Damietta branch, and cover a small area of 6.83 km^2, or 0.05% of the total area of the fourth time deposits, as shown in Table 2.1 and Fig. 2.4.

2.2.3.3 Stabilized Dunes

Old fixed dunes are characterized by not exceeding 5 m above sea level with a scarcity of vegetation, consisting of very fine sand and silt, and are associated with the first generation of the sand transmission system by winds from the old estuary, which has been subjected to erosion (Spinite) since 2000, approximately 2500–2000 BC (El-Asmar and Al-Olayan 2013). The fixed ancient dune deposits occupy an area of 869.38 km^2, or 6.36% of the total area of the fourth time sediments, and are distributed in several locations in the study area (south of Idku Lake, south of Rashid city, Ras Abu Qir, south of Burullus Lake, south of Manzala Lake, Manzala Lake Barrier, and on both sides of the navigational course of the Suez Canal, North Sinai coast on the eastern side of the Suez Canal), as shown in Fig. 2.4.

2.2.3.4 Coastal Dunes

They are young, active, fine sand, Holocene sand dunes, whose ranges are concentrated along the coast of the Nile Delta and extend through the backshore up to 10 km south. There are two distinct types of modern sand dunes along the Nile Delta coast, namely, barchans and longitudinal dunes. It is clear from the analysis of the geological map and field studies that the sand dunes in the study area are distributed in several fields (Rasheed field, Idku, Al-Khashaba field east of the mouth of Rashid, Baltim field, Abu Madi, Gamasa field, Damietta, a field in the middle of the marshes on the eastern side of the Suez Canal, and the area of dune sediments is Sandy 517.13 km^2, or 3.78% of the total area of the fourth time sediments, as shown in Fig. 2.5.

2.2.3.5 Sabkhas and Salt Marches

Sabkhas are defined as low deflation surfaces that were formed by removing the loose dry materials they contain until they reached the groundwater level and are characterized by the deposition of Evaporate Minerals, gypsum, and anhydrite and are characterized by the presence of temporary depressions that are submerged in water to form playa lakes and then evaporation works to dry the lakes again and sedimentation of suspended and dissolved loads (El-Banna and Frihy 2009). The study area includes salt marshes, saltpans, wetlands around coastal lagoons, and fish farms. Its distribution is concentrated on a large scale in the study area, especially south of Manzala Lake and on the western and southern sides of Burullus Lake, Mariout, and Rashid Estuary. The sabkhas occupy an area of 1510.16 km^2, or 11.05% of the total area of the fourth time sediments, as shown in Fig. 2.4, Table 2.1, and Fig. 2.5.

2.2.3.6 Quaternary Deposits (Q)

Undifferentiated these deposits belong to the fourth time (Pleistocene) and consist of Duricrusts, sand, pebbles, and recent coastal sediments. They cover an area of 671.06 km^2, or 4.91% of the total area of the fourth time deposits, as shown in Table 2.1. These sediments are generally distributed over most of the study area, as they cover the western side of the Nile Delta, the area around the Rosetta estuary, the Burullus Lake barrier area, and most of the coast between Baltim and the Damietta estuary, as shown in Fig. 2.4. These sediments are classified as alluvial to shallow marine clastic sands, tend from brown to light olive gray to olive gray, and are characterized by fine-grained to coarse sands that are poorly sorted and appear on the eastern side of the study area. These sands consist of iron-stained quartz, feldspar, mica, heavy metals, and crustaceans (Warne and Stanley 1993).

2.2.3.7 Coastal Ridges (CR)

On the western side of the study area, sediments of the coastal margins are distributed in the area between Burj Al-Arab and Al-Omayed (El-Asmar and Wood 2000, p. 1138). Among the most important edges extending from the study area, the initial coastal edge, consisting mainly of solid white loess limestone, the most widespread, the edge of Al-Max-Abusir, composed of solid white limestone and chalk, and the grains of limestone that make up this series are generally smaller in size than those that make up the coastal series, and the edge of Mount Mariout, and this series consists of Sandy limestone, the chain is 35 m high, and the limestone rocks vary greatly in their grain sizes, ranging from 0.5 mm to 1.4 mm. The sandy limestone cover is solid dark brown, and this chain is rich in calcareous algae and Mediterranean foraminifera. These foramans date back to the Pleistocene era (Attia 1975), as shown in Fig. 2.4. These deposits occupy an area of 44.16 km^2, or 0.32% of the total area of the fourth time deposits.

2.2.3.8 Calcarenite Bars of Pleistocene Age (Qc)

They are rocky calcareous ridges consisting of chalky limestone, marl, and sandy limestone followed by carbonate ridges belonging to the Alexandria Formation. According to the geological maps of the study area, it constitutes an area of 73.69 km^2, or 0.54% of the total area of the fourth time sediments, as shown in Fig. 2.4 and Table 2.1. It spreads only on the northwestern side of the study area, especially in the area confined between Alexandria and Al-Hammam. The coast in this range is distinguished by its white color and carbon beaches, in addition to Lake Mariout and shallow water surfaces that are connected to the sea only through a number of artificial canals and polluted by industrial waste and sewage and industrial wastes. It is also characterized by the presence of some carbon edges. Parallel, which is slightly more than 40 m high, separated by shallow depressions and sabkhas, and a wide section of this area, is located below sea level (Warne and Stanley 1993).

2.2.4 Geological Structure of the Nile Delta

Geologically, the greater Nile Delta region covers the following structural-sedimentary provinces: (1) the south delta block, (2) the Nile cone, and (3) the Levant platform (Sestini 1989). The hinge zone: The boundary between the south delta block and the north delta basin is a major zone of flexure at approximately 31 °N latitude, along which faults downthrown to the N by 5000–6000 m into the Cretaceous-Middle Eocene carbonates of N Egypt (Sestini 1989), as shown in Fig. 2.6.

The hinge zone is a faulted flexure zone 30–40 km wide. Its age was dated back to a Jurassic crustal break, representing the boundary between a southern stable platform (South Delta block) and a northern subsided basin, where all Cenozoic sequences present thicker and relatively deep marine successions (Kamel et al. 1998). The hinge

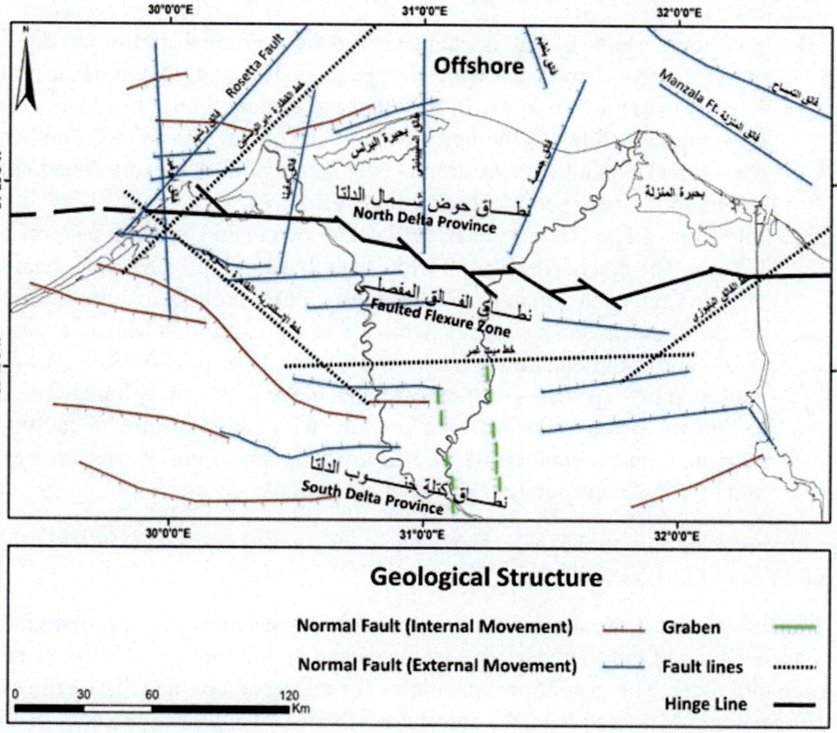

Fig. 2.6 Main subsurface structures of the Nile Delta region (*Source* After [Barakat 2010; Sestini 1989])

zone has played a dominant role in the stratigraphic and tectonic evolution of the Nile Delta (Said 1981). South of the Flexure zone, the principal pre-Oligocene structures are tilted fault blocks and horsts oriented E–W to NNE; in the north, the Sinai-Negev equivalent structures and surface folds, onshore and offshore, are oriented NE (Sestini 1989).

2.2.5 The Nile Delta is Differentiated into Two Geological Provinces

1. The deep offshore Nile Delta (north of the continental shelf, i.e., north of the 200 m isobath and west of a principal NE–SW Pliocene fault, which was affected by strong Pliocene–Pleistocene sediment loading (3500 m in 4–5 m/year). It displays large-scale postMessinian listric faulting, marked rollovers, rotated blocks, and slump structures, especially in the NNE and NE (Sestini 1995).
2. Onshore Nile Delta: The onshore Nile Delta region is divided by the flexure zone, which is known as a hinge line, into two structural-sedimentary sub provinces: the South Nile Delta block and the North Nile Delta basin.

    2.1. The South Delta Block: is characterized by a gradual northward dip of middle Eocene carbonates, which is represented by gently asymmetric folds referred to as the Syrian Arc fold system and extends along an arcuate trend from northern Sinai to the north of the Gulf of Suez across the southern part of the Delta into the Western Desert. Some of these flexure zone faults extend to Pliocene sediments (Kamel et al. 1998).

    2.2. The North Delta Basin: is characterized by two main structural patterns as follows: The first is a deep pre-Tortonian fault pattern (possibly to Eocene— or Late Cretaceous) mainly of E–W fault blocks, prominent among which are the shelf margin structures, which play a great role in Miocene subsidence and sedimentation. The second is a shallow post–Messinian fault pattern; these faults are genetically related to the sedimentary load of recent sediments at the unstable Delta margin, which caused growth faulting, slumping, and normal faults as well as diapirism of uncompact Pliocene and Messinian evaporates (Kamel et al. 1998).

The structural trends of the Nile Delta were recognized by Sestini (1995) and Zaghloul et al. (2001) as follows:

1. The Tethyan trend, an east–west trend, could be related to the original continental margin rifting of the southeastern Mediterranean during the early Mesozoic and probably older. The best-known examples for this trend are the Oligo-Miocene Hinge Zone, Mit-Ghamr Fault, and the northern and southern flexures of the onshore Nile Delta.
2. The Rosetta trend, a northeast–southwest trend Late Cretaceous age, is exemplified by the Pelusium, Qattara–Eratosthenes, and the Gamasa, Idfina, and Port

Said-fault lines. The faults are likely to have originated from one point in the northeast corner of the Mediterranean Sea at Alexandria. In addition to the vertical motion component of these faults, they exhibit sinistral strike-slip displacement.
3. The northwest–southeast trend was active during the Miocene. Its best-known example is the Temsah or Bardawil Line in the eastern offshore Nile Delta. The majority of the faults

2.3 Morphological Characteristics

The morphological study is a very important factor in the assessment and analysis of the impacts of potential geohazards related to sea level rise on the coastal low-laying zone. The morphological setting of the Nile Delta coast was divided into three parts: (1) topographic analysis, (2) slopes, and (3) aspects.

2.3.1 Topographical Analysis

The Nile Delta in Egypt is one of the earliest recognized deltaic systems in the world. The Greek historian Herodotus in 450 BC was the first to describe its triangular shape as "Delta" since it resembles the inverted Greek letter Δ. It was formed by sedimentary processes between the upper Miocene and present (Nelsen 1976; Stanley and Warne 1993) and built up by the alluvium brought by the old seven active branches of the Nile. Those distributaries have been subsequently silted up and replaced by the present Damietta and Rosetta branches (El-Banna and Frihy 2009).

The shoreline of the Nile Delta, from El Hamam to AbuQuir, consists of a carbonate ridge, and from AbuQuir to the east of Port Said, 275 km long, is a gently arcuate coast consisting of protruding promontories separated by embayments and saddles. Delta beaches and their contiguous coastal flats are partially backed by coastal flats, dunes, or brackish lagoons. Three lagoons (Idku, Brullus, and Manzala) are separated from the sea by narrow barriers, leaving artificial inlets.

The global Digital Elevation Models of the USGS seamless Shuttle Radar Topography Mission (SRTM, 3Arc ver.2, 90 m resolution) taken in 2012 and covering the Nile Delta in Egypt in 9 scenes were collected and preprocessed for use in the topographical analysis of the Nile Delta coast in several steps using a different software: Global Mapper v. 2016, ERDAS Imagine v.2014, and ArcGIS v.10.2: Color Matching and histogram equalizations, Mosaic, reprojection to different metric projection (UTM, WGS, Zone 36 N), fill null value gaps, Generalization using the Focal Statistics 11 × 11 mean value, and reclassification of the raw elevation data into specific elevation zones, as shown in Fig. 2.7.

It is clearly seen that the topography of the lower Nile Delta has a large area between zero and 1 m elevation, with parts below sea level −6 m. Areas below mean sea level include the coastal lagoons, the former AbuQuir/Maryut lagoon south

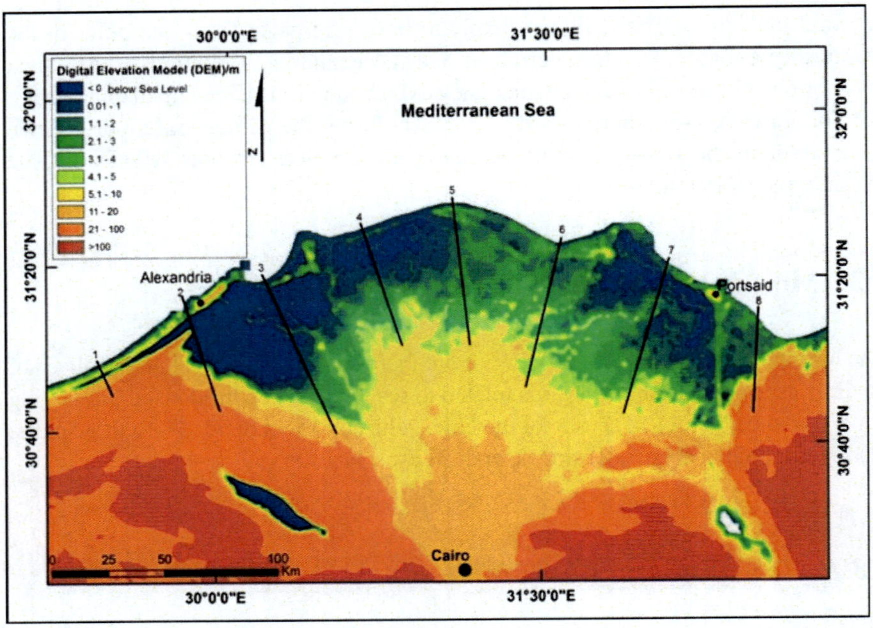

Fig. 2.7 Topography of the Nile Delta coast in Egypt (SRTM, 3Arc ver. 2, 2012, 90 m)

of Alexandria, and the aquacultures bordering the southern margins of the coastal lagoon. Delta lagoons form a transition zone between land and the sea, in most places separated from the sea by narrow and low-lying sand barriers. A great part of the beach and coastal flat lies between zero and 2 m above mean sea level. Areas above 3 and 4 m lie within the coastal dunes at the backshore of Abu Quir Bay, Gamasa, and in the southern part of the lower delta plain, corresponding to approximately 35 km from the shoreline.

Eight longitudinal topographical profiles have been measured from the digital elevation model from the sea to the inland to analyze the topography of the Nile Delta coast; they cross the Nile Delta coast in different locations: western, central, and on the eastern Nile Delta coast (Figs. 2.7 and 2.8).

2.3.1.1 Topographic Description of the Nile Delta Coast

According to Fig. 2.8, the first profile was measured near Hamam city at the western part of the Nile Delta coast, directed NW–SE, and approximately 16 km in length. The coastal plain at this part is too narrow and distinctive with a carbonate ridge (7 m height and 3 km width) separating the Maryut lagoon (−3 m) off the Mediterranean Sea, then it lies in front of a very high elevation limestone plateau graduates in elevation to 60 m above the sea level toward the south.

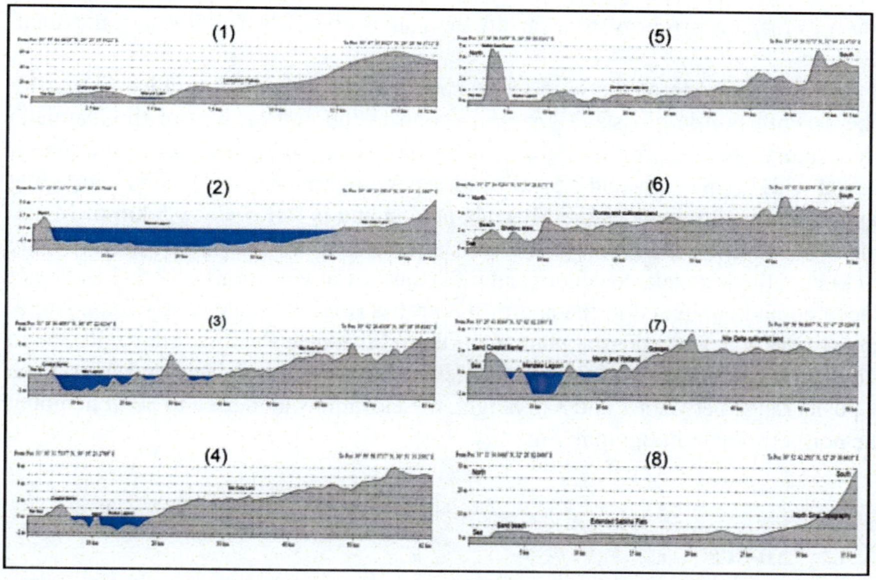

Fig. 2.8 Topographical profiles along the Nile Delta Coast

The second profile shown in Fig. 2.8 describes the area behind Alexandria city. This section is directed NS-SE and 54 km in length. According to the topographic profile, a coastal barrier approximately 2 m in height and 2 km in length separates the below sea level land at Maryut Lagoon off the sea, while the elevation in the southwestern part of this section tends to increase toward the Nile Delta floor.

The third profile crossing AbuQuir Bay eastern Alexandria city, Its directed NE–SW with 83 km length, there is a very small coastal barrier 0.7 m height and less than 1000 m in width separates the Idku lagoon off the Mediterranean sea, this low-laying lands around the Idku lagoon continue up to 40 km along the profile, then the elevations start increasing to 6 m height southern Idku lagoons shown in (Fig. 2.8).

Three profiles represent the central Nile Delta zone (4, 5, and 6). The fourth profile crossing El- Brullus lagoon at the eastern of Rosetta promontory, it is approximately 63 km length directed NW–SE, a small coastal barrier 1.2 m height and 5 km width separate a low-laying lands 10 km width along the profile include the Brullus lagoon off the Mediterranean Sea, then the elevations start to increase to the Nile Delta floor to 6 m height southern the Brullus lagoon shown in (Fig. 2.8).

The fifth profile crosses the Balteam coast near the central line of the Nile Delta coast. This profile is 48.5 km in length and directed N–S. A high elevation 5 m coastal barrier 3 km in width consists of an Aeolian Sand Dunes separating the Brullus lagoon to the south approximately 25 km in width. Then, the profile's elevation increased gradually to 5 m height and 45 km south of the Balteam coast, as shown in Fig. 10.5. The sixth profile crosses Gamasa city to the southwest with a width of 51 km, and this profile shows an area that increases gradually in elevation from the north to the south to reach a height of 6 m far off the gamasa beach with 50 km in the south (Fig. 2.8).

The eastern zone of the Nile Delta coast is represented by profiles (7 and 8). The seventh profile crosses Manzala Lagoon in the eastern part of the Damietta promontory, and this profile is 58 km wide from NE to SW. A stabilized dune coastal barrier with 1.5 m height and 2.5 km width separating the Manzala lagoon low-laying land 12 km width includes the Manzala lagoon and is followed by 5 km of March and wetland. Then, the elevation increased dramatically to 4 m height approximately 30 km off the Manzala coastline, and a flat surface of elevations of 2.5–3 m height appeared and continued to the end of the profile (Fig. 2.7). The last profile, No. 8, crossed the Nile Delta coastal plain in northern Sinai near Palouse village. Its 35 km width and most of the profile lay in low-elevation coastal zone beach sand and sabkha deposits range between 1 and 3 m height, the elevations increased to 30 m height at the northern Sinai topography Fig. 2.8.

2.3.2 *Slopes*

It is obviously clear from the analysis that the slopes map for the study area (Fig. 2.9). Most of the Nile Delta land did not exceed a slope degree of 0.2, except for the coastal plain at the western portion of the Nile Delta, which had a slope degree exceeding 1. On the other hand, the western and eastern Platueas surrounding the Nile Delta have higher degrees of slopes exceeding 10 degrees at the western Platuea and 5 degrees at the eastern Platuea.

2.3.3 *Aspects*

It is clearly seen from the analysis of the aspects along the Nile Delta coast and the surrounding area, as shown in Fig. 2.9. The dominant slope directions along the Nile Delta coast are toward the north, northeast, and northwest according to the degree angles shown in Fig. 2.9. The directions of slopes follow the local landforms distributed along the Nile Delta floor, such as the Lagoons, ridges, depression, and lakes. Slopes and their aspects are very important in the analysis of sea level rise and its potential geohazards along the coastal plain of the delta.

2.3.4 *Climatological Characteristics*

The climatological norms published by the Egyptian Metrological Authority between 1942–1975 and 1976–2005 are used to analyze the effective climatological elements at five different stations along the Nile Delta coast (Alexandria Alnosha, Rosetta,

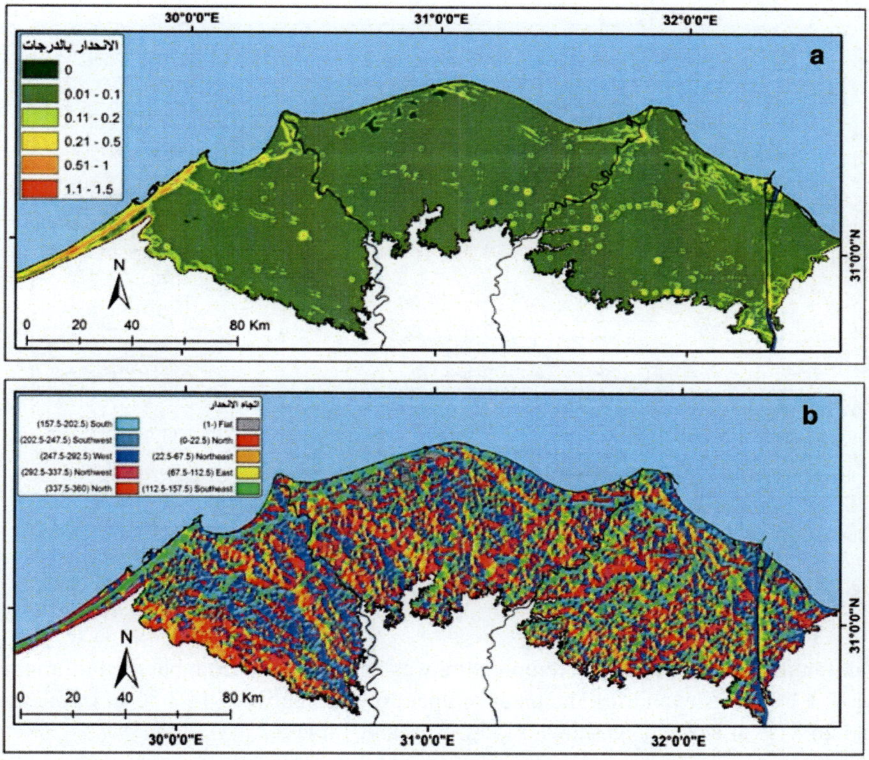

Fig. 2.9 Map of slope and aspect along the Nile Delta coast

Balteam, Damietta, and Port Said). The effective climate elements studied in this part (temperature, relative humidity, rainfall, and wind speed) were analyzed in the period from 1942 to 2005.

2.3.5 Temperature

The temperature trends along the Nile Delta coast were analyzed monthly. (1) The winter: the maximum temperature was 19.9 °C at Port Said and Alexandria in December, and the lowest temperature was 8.4 °C in January at Damietta. The mean temperature in the winter was confined between 12.8 °C at Damietta and 16.6 °C at Rosetta, as shown in Fig. 2.10. (2) Spring, the maximum temperature was 26.6 °C at Damietta, but the minimum was 11 °C at Alexandria. The mean temperature lies between 15.6 °C at Alex and 23 °C at Rosetta. (3) In the summer, the maximum temperature was 30.9 °C at Port Said, but the minimum was 19.8 °C at Damietta. The mean temperature was between 24.3 °C at Balteam and 28.1 °C at Rosetta. (4)

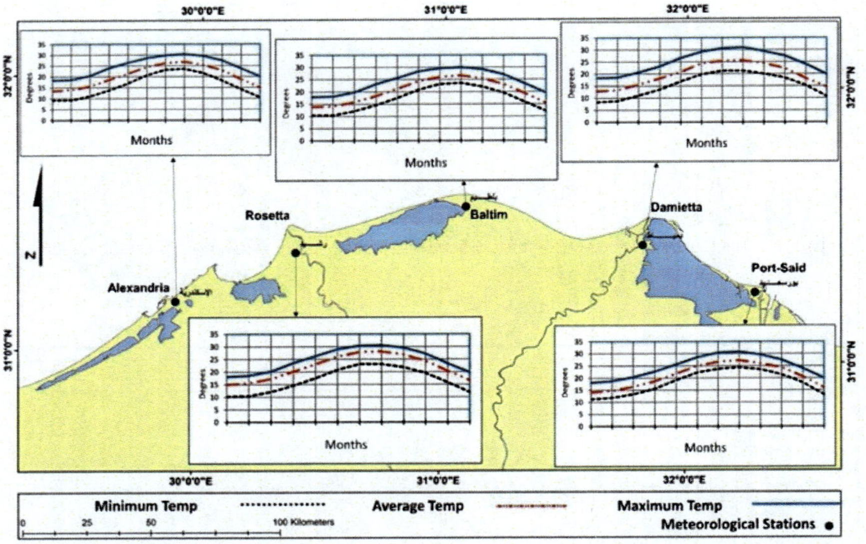

Fig. 2.10 Monthly mean temperature at the Nile Delta coastal stations from 1942 to 2005

Fall (autumn): The maximum temperature was 29.6 °C at Rosetta, but the minimum was 14.1 °C at Alexandria. The mean temperature lies between 18.1 °C at Damietta and 26.8 °C at Rosetta, as shown in Fig. 2.10 and Table 2.2.

2.3.6 *Relative Humidity*

The analysis of the relative humidity along the Nile Delta coast at five different stations is shown in Table 2.3. (1) In the winter months, the lowest relative humidity percent was 65.1%, which was recorded at the Rosetta station in February. The highest percent was 75% at the Damietta station in January. (2) In the spring months, the lowest percent was 63.2% at Rosetta station, and the highest percent was 70.5% at Balteam station. (3) In the summer months, the lowest percent humidity was 64.2%, recorded at Rosetta station, and the highest was 73.5% at Balteam station in the middle delta. (4) In the fall months, the lowest percent was 65% in Rosetta during September, and the highest was 75% in Damietta during September and November. Generally, the highest humidity percentages throughout the year were recorded in the fall and winter months, as shown in Table 2.3.

Table 2.2 Temperature (C°) recorded at the Nile Delta coast (1976–2005)

	Alexandria St			Rosetta St			Balteam St			Damietta St			Port Said St		
	Min	Mean	Max	Min	Mean	Max	Min	Mean	Max	Min	Mean	Max	Min	Mean	Max
January	9.1	13.6	18.3	10.2	15.0	18.1	10.4	13.9	17.9	8.4	12.8	18.3	11.2	14.2	18.1
February	9.2	13.9	18.6	10.4	15.5	18.5	10.4	14.1	18.1	8.8	13.4	18.6	11.8	14.7	18.8
March	11	15.6	20.4	12.0	17.2	20.1	12.0	15.7	19.7	11.1	15.6	20.5	13.3	16.4	20.4
April	13.7	18.6	24.0	14.5	20.1	23.6	14.6	18.4	23.0	13.6	18.0	23.1	15.6	18.7	22.6
May	16.9	21.4	26.4	17.5	23.0	26.2	17.4	21.2	25.7	16.8	20.9	26.6	19.1	21.8	25.7
June	20.7	24.4	28.5	20.9	26.1	28.9	20.9	24.3	28.3	19.8	24.4	24.2	22.2	25.0	28.6
July	23.1	26.3	29.8	23.0	27.8	30.4	23.1	26.1	29.6	21.2	25.4	30.6	23.8	26.6	30.4
August	23.6	26.8	30.3	23.1	28.1	30.7	23.5	26.6	30.0	21.2	25.7	31.0	24.4	27.3	30.9
September	21.6	25.3	29.6	21.9	26.8	29.6	22.0	25.4	29.2	20.0	24.3	29.4	23.5	26.1	29.5
October	18.2	22.8	27.4	19.5	24.3	27.2	19.5	23.0	27.1	18.4	22.2	27.4	21.4	24.3	27.4
November	14.1	18.8	23.7	15.8	20.4	23.4	15.9	19.3	23.4	15.2	18.1	23.9	18.0	20.6	23.9
December	10.3	15.0	19.9	11.8	16.6	19.7	12.0	15.4	19.6	10.6	14.5	19.8	13.1	16.0	19.9
Annual mean	16	20.2	24.7	16.7	21.7	24.7	16.8	20.1	24.3	15.4	19.6	24.9	17.8	21.1	25.4

Source Climatological Normals between 1976 and 2005, Published by the Egyptian Metrological Authority

Table 2.3 Climate data of humidity and rainfall along the Nile Delta coast

Month Stations	Alexandria St		Rosetta St		Balteam St		Damietta St		Port Said St	
	RH	R	RH	R	RH	R	RH	R	RH	R
January	70.2	53.1	67.5	52.9	72.4	50.4	75	25.5	71	13.5
February	68.2	14.7	65.1	32.6	70.6	40.2	72	17.2	68	11.7
March	67.6	14.6	64.3	19.9	70.5	17.2	70	10.7	66	8.8
April	65.4	3.2	63.2	6.1	68.6	5.3	69	3.7	69	3.7
May	67.3	1.7		2.5	69.8	4.3	68	1.9	69	2.2
June	69.3	0.1	64.2	0.2	71.5	0.3	70	0.1	70	Tr
July	71.0	2.0	66.8	0.0	73.5	0.0	70	Tr	71	0.0
August	70.7	0.0	67.1	0.0	73.2	0.0	76	Tr	71	Tr
September	67.8	2.7	65.0	5.9	70.1	4.7	75	0.5	68	5.4
October	68	9.9	65.6	9.8	69.8	9.5	74	7.1	68	19.6
November	69	28.5	66.9	36.3	70.5	23.5	75	15.4	70	18.0
December	70.5	49	69.5	57.3	72.8	38.7	74	24.6	71	47.7
Annual mean	68.8		63.7		71.1		72		68	

(RH) = Absolute Humidity %, (R) = Rainfall mms
Source Climatological Normals between 1975 and 2005, Published by the Egyptian Metrological Authority

2.3.7 Precipitation (Rainfall)

The analysis of the recorded statistics for the precipitation along the Nile Delta coast at five different stations refers to the following: (1) In the winter, 11.7 mm was the minimum amount registered at Port Said station in February, while the highest amount was 57.3 mm at Alexandria station in December. (2) In the spring months, the minimum amount was 1.7 mm at Alexandria, but the maximum amount was 19.9 mm at Rosetta. (3) The summer months are considered the lowest months among the year in terms of precipitation, with a maximum amount of 2.0 mm and a minimum amount of 0.0 mm at the stations of study. (4) In the fall months, the maximum amount of precipitation was 36.3 mm recorded at the Rosetta station, and the minimum was 0.5 m at the Damietta station, as shown in Table 2.3 and Fig. 2.11.

2.3.8 Wind

The analysis of the wind speed along the Nile Delta coast indicates that Port Said and Alexandria have the highest annual mean wind speeds of 9.0 and 7.9 knots,

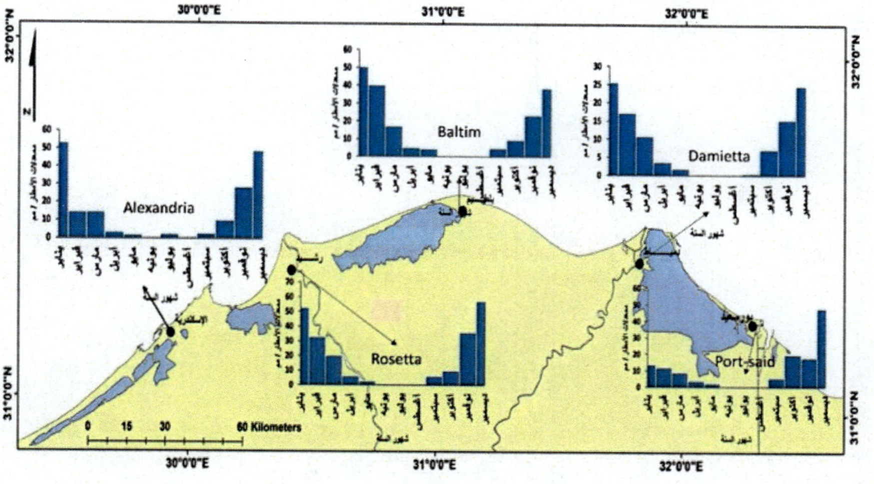

Fig. 2.11 Rainfall annual rate in the Nile Delta coastal stations

respectively. The other three stations have wind speeds lower than 5.6, 5.5, and 5.4 knots at Damietta, Balteam, and Rosetta, respectively, as shown in Table 2.4.

The monthly analysis was analyzed according to Table 4.3, and the results were as follows: (1) in the winter, the lowest wind speed was 4.7 knots, which was registered in Balteam during the December month, but the highest wind speed was 9.3 knots registered in Port Said during January. (2) In the spring months, the lowest wind speed was 5.6 knots registered in Balteam during May, and the highest wind speed was 11.3 knots at the Port Said station during March. (3) During the summer months, the lowest wind speed was 4.8 knot registered in Damietta during August, while the highest was 9 knots at Port Said during June. (4) The highest wind speed in the fall months was 8.3 knots, which was registered at Port Said station during November, and the lowest wind speed was 4.4 knots at Damietta station during September, as clearly seen in Fig. 2.12 and Table 2.4.

2.4 Summary

This chapter addresses the study of the natural characteristics of the study area, because of its paramount importance to identify the nature of the study area and the geographical factors affecting its dynamic. This chapter was divided into three main sections: The first section examines the geological characteristics of the study area through the study of the formation and evolution of the Nile Delta, the stratigraphic succession of subsurface formations, the geographical distribution of surface geological deposits, and their presence ratios. The second section studied the morphological

Table 2.4 Wind speed (knots) along the Nile Delta coast

	January	February	March	April	May	June	July	August	September	October	November	December	Annual mean
Alexandria St	7.8	8.4	8.6	8.4	8.0	8.4	8.9	8.6	7.4	6.8	6.8	7.2	7.9
Rosetta St	5.1	5.8	6.0	6.0	5.8	5.8	5.8	5.3	4.9	4.7	4.9	4.9	5.4
Balteam St	5.3	5.8	6.2	6.2	5.6	5.8	6.6	5.8	4.7	4.5	4.5	4.7	5.5
Damietta St	5.5	6.0	7.2	6.9	6.1	6.1	5.4	4.8	4.4	4.8	4.8	5.6	5.6
Port Said St	9.3	10.0	11.3	10.4	9.4	9.0	8.4	7.4	7.4	8.0	8.3	8.4	9.0

Source Climatological Normals between 1975 and 2005, Published by the Egyptian Metrological Authority

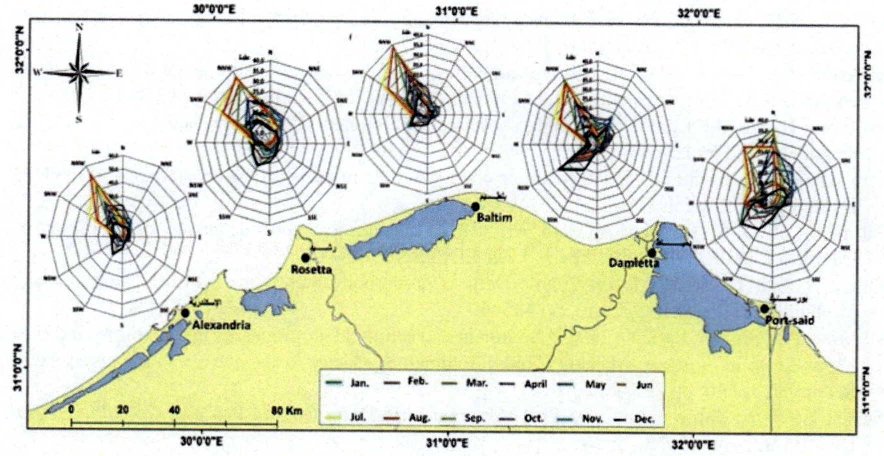

Fig. 2.12 Monthly wind speed at the Nile Delta coastal stations from 1942 to 2005

characteristics of the study area and included the study of the topographic characteristics through the analysis of the digital elevation model prepared for the study area, including the characteristics of the terrain ranges and the terrain sectors. It was clear from the analysis that a large part of the study area fell below sea level, especially the subtraction area, which represents the old Abu Qir lake and the depression of the old Lake Mariout, and then the degree of slope varied depending on the local terrain of sand dunes, carbon ridges, and beach terraces. The last section addresses the study of the climatic characteristics of the study area, including in detail the study of the climatic elements represented in temperature, wind, rainfall, hours of solar brightness, relative humidity, evaporation, and the extent of their impact on the study area.

References

Abou El-Magd I, Hermas EA (2010) Human impact on the coastal landforms in the area between Gamasa and Kitchener drains, Northern Nile Delta, Egypt. J Coastal Res 26(3):541–548

Allen PA (1997) Earth surface process. Blackwell, Dublin, 416p

Attia SH (1975) Petrology and soil genesis of the quaternary deposits in the region west of the Nile Delta (North and East of Wadi EL Natrun). Ph.D. Thesis, Faculty of Science, Ain Shams University, Cairo

Barakat MK (2010) Modern geophysical techniques for constructing a 3D geological model on the Nile Delta, Egypt. A Ph.D. Dissertation, Technischen Universität Berlin, Berlin

Bartolini C, Malesani PG, Manetti P, Wezel FC (1975) Sedimentation and petrology of quaternary sediments from the Hellenic Trench, Mediterranean Ridge and Nile Cone, DSDP Leg 13 cores. Sedimentology 22:205–236

Chen Z, Warne AG, Stanley DJ (1992) Late quaternary evolution of the northwest Nile delta between Rosetta and Alexandria, Egypt. J Coast Res 8:527–561

El Banna MM, Frihy OE (2009) Human-induced changes in the geomorphology of the northeastern coast of the Nile delta, Egypt. Geomorphology 107:72–78
El-Asmar H, Al-Olayan H (2013) Environmental impact assessment and change detection of the coastal desert along the central Nile Delta coast, Egypt. Int J Rem Sens Appl 3:1–12
El-Asmar H, Wood P (2000) Quaternary shoreline development: the northwestern coast of Egypt. Quat Sci Rev 19:1137e1149
El-Nahry AH, Doluschitz R (2010) Climate change and its impacts on the coastal zone of the Nile Delta, Egypt. Environ Earth Sci 59:1497–1506
Farouk S, Ziko A, Eweda SA, Said AE (2014) Subsurface Miocene sequence stratigraphic framework in the Nile Delta, Egypt. J Afr Earth Sci 91:89–109
Kamel H, Eita T, Sarhan M (1998) Nile Delta hydrocarbon potentialities. 14th Exploration and production Conf., EGPC, Cairo, pp 485–503
Milliman JD, Qin YS, Park YA (1989) Sediment and sedimentary processes in the Yellow and East China Seas. In: Taira A, Masuda F (eds) Sedimentary facies in the active plate margin. Terra Scientific, Tokyo, pp 233–249
Nelsen E (1976) Shore evolutions. Proc. Seminar Nile Delta Coastal Process, CoRI/UNESCO/UNDP Cairo, pp 15–59
Rizzini A, Vezzani F, Cococcetta V, Milad G (1978) Stratigraphy and sedimentation of a Neogene-Quaternary section in the Nile Delta area (A.R.E.). Mar Geol 27:327–348
Said R (1981) The geological evolution of the river Nile. Springer-Verlag, New York, p 151p
Said R (1990) The geology of Egypt. Balkema, A.A
Said R (1993) The river Nile: geology, hydrology, and utilization. Pergamon Press
Sestini G (1989) Nile Delta: a review of depositional environments and geological history. In: Whateley MKG, Pickering KT (eds) Deltas: sites and traps for fossil fuels. Geol Soc London Spec Publ. 41:99–127
Sestini G (1995) Egypt. In: Kulke H (ed) Regional petroleum geology of the world, part II: Africa, America, Australia and Antarctica (Beiträge zur regionalen Geologie der Erde 22: 66–87, Gebrüder Borntraeger Verlagsbuchhandlung, Stuttgart)
Stanley DJ, Warne AG (1993) Nile delta: recent geological evolution and human impacts. Sci J 260:628–634
Stanley DJ, Warne AG (1994) Worldwide initiation of Holocene marine deltas by deceleration of sea-level rise. Science 265:228–231
Stanley DJ, Warne AG (1998) Nile delta in its destruction phase. J Coastal Res 14:794–825
Warne AG, Stanley DJ (1993) Archaeology to refine Holocene subsidence rates along the Nile delta margin. Egypt Geology 21:715–718
Zaghloul ZM, Elgamal MM, Shaaban FF, Yossef AFA (2001). Plates interactions and petroleum potentials in the Nile Delta. In: Zaghloul Z, El-Gamal M (eds) Deltas (ancient and modern), pp 41–53
Zuidam V, Robert A, Farifteh J, Marieke A, Cheng T (1998) Developments in remote sensing, dynamic modeling and GIS applications for integrated coastal zone management. J Coastal Conser 4(2):191–202

Chapter 3
Assessment of the Nile Delta's Coastline Dynamics: A Remote Sensing and GIS-Based Computational Approach

Abstract Delta coastlines are considered one of the most morphodynamic and vulnerable landforms on the earth's surface. Natural and anthropogenic factors affected the behavior and trend of coastline retreat and advance along the NDC. Previous studies indicated that marine climate, sea level rise, and dams along the lower part of the river were the main factors of Nile Delta coastline deformation. In the last decade, geospatial technologies have been widely utilized for coastline change analysis and mapping along different coasts worldwide. This chapter aims to use multitemporal satellite remote sensing data, historical topographic maps, and GIS-based spatial computational techniques for historical coastline dynamic assessment along the NDC from 1945 to 2015. The Digital Shoreline Analysis System (DSAS) developed by the USGS and working in the ESRI ArcGIS environment was applied to assess the spatiotemporal variation in coastline dynamics. Four different statistical parameters were applied in this study; the results were assessed and mapped using GIS techniques.

Keywords Coastline dynamics · DSAS · Nile Delta coast · Shoreline retreat

3.1 General

Deformation of coastlines and related hazards have become important environmental problems for planning and sustainable development. Previous studies have identified several geomorphological, sedimentological, and archeological indicators for coastline retreat along the NDC.

This chapter aims to assess coastline dynamics along the NDC during the period (1945–2015). The quantitative assessment was performed based on multisource geospatial data, including the Egyptian topographic maps (scale 25,000) produced by the Egyptian Surving Authority in 1945 and multisensor Landsat imagery taken from 1972 to 2015. Spectral indices such as NDWI and MNDWI were applied using ERDAS IMAGINE v.2014 Model Maker to extract multiple coastlines from Landsat imagery over time. The Digital Shoreline Analysis System (DSAS) version 4.3 is

K. S. Darwish, *Hazard Modeling and Assessment of the Nile Delta Coast*,
https://doi.org/10.1007/978-3-031-44324-4_3

an extension of ESRI ArcGIS v.10.2 software that was produced by the USGS and used to calculate shoreline rate-of-change statistics. A GIS geodatabase was built, including the appended shoreline layer, baseline layer, and Transects layer, which were automatically generated. Automatic computerized quantification processing was performed on the DSAS database to calculate the *net shoreline movement* (NSM), *shoreline change envelope* (SCE), *end point rate* (EPR), and linear regression rate (LRR). All of the results were spatially mapped and analyzed. The spatiotemporal coastline change analysis selected four periods for study: the first period before the construction of the High Dam in 1964 (1945–1972), the second period after the construction of the Aswan High Dam (1972–1984), the third period before the construction/during the construction of coastal protection devices (1984–2001), and the last period after the construction of coastal protection (2001–2015).

3.2 Factors Controlling Coastline Dynamics

There are many factors controlling coastline changes, sediment movement and erosion and sedimentation processes, the most important of which are as follows:

3.2.1 Hydrodynamic Factors

These factors are represented in the marine climate of waves, winds, tides, and shore currents. They contributed to the formation, change, and evolution of coastline morphology.

3.2.1.1 Wave Action

A NDC is defined as a coast typically dominated by waves or current-dominated waves. Statistical analysis of the wave regime records measured in AbuQir Bay and Damietta Port shows that the prevailing waves are mainly concentrated in two quadrants: (north–west) and (north–east) (Frihy et al. 2003). Northwest waves have a great impact on morphological processes due to their long period, especially in the winter, and are responsible for generating net sediment movement along the NDC. The waves coming from the north–east direction prevail seasonally from (4 to 6 months) during the periods of studying the waves in the areas of Rosetta and Damietta, and the south–west direction extended only for a period ranging between (1 and 3 months) of the months of the study.

It is clear from the analysis of the monthly distribution of wave directions that the waves blow from two main directions: a north–west direction that includes (north–northwest, northwest, and north–west) and a north–east direction that includes (north–northeast, northeast, east, and northeast). Measurements at the AbuQir station indicated that the north–west direction occupies 81%, the north–east direction occupies 14%, and from the south–west direction, waves blow by 5%.

According to Frihy et al. (2003), the highest height of the significant waves west of the Rosetta estuary was 5.4 m from the west–northwest direction in December 1988, and the average wave height was 1.2 m at a frequency of 5.6 s in 13 months, as evidenced by the wave directions rising (37-a). The highest height of the waves affecting the port of Damietta was 4.2 m from the north direction during January 1998, and in general, the average wave height and frequency reached 0.5 m and 6.3 s, respectively. In addition, Iskander (2013) indicated that there was an increase in the average height of the significant waves during the period between 1985 and 2010 at a rate of 2.6–2.9 cm/year. The study confirms that the wave energy in front of coastal protection means is expected to increase by 20% over the next 50 years with high storms because of global climate change.

3.2.1.2 Wind Action

The analysis of wind speed/direction frequency at three stations (Alexandria, Rosetta, and Baltim) during 1976–2005. The northwest winds are the strongest prevailing winds that affect the movement of waves in the study area, which have a speed of more than 11 knots at Alexandria station from directions (north–northwest, north–west, and north, with percentages of 28.2%, 15.4%, and 13.5%, respectively). At Rosetta station, winds blow from north–northwest, north–west, north, and northeast directions by 23%, 18.2%, 12.5%, and 9.4%, respectively. At the Baltim station, winds blow from north–northwest, north–west, north–west, and westerly directions at rates of 24%, 14%, 7.2%, and 6.7%, respectively.

Calm winds at Alexandria station occupy 18.2% with speed (1–3 knots), winds with speed (4–6 knots) occupy 21.6%, winds between (7–10 knots) occupy 31.9%, winds ranging from 11–16 knots occupy 23.7%, and winds with high speed of more than 17 knots occupy 4.6%. At the Rosetta station, calm winds occupy 44% with velocities of 1–3 knots, winds with speeds of 4–6 knots occupy 33.1%, winds with speeds of 7–10 knots occupy 16.4%, winds with speeds of 11–16 knots occupy 4.8%, and high-speed winds of more than 17 knots occupy 1%. At the Baltim station, wind with a speed of 1–3 knots occupies 4.4%, wind with a speed of 4–6 knots occupies 28.9%, wind with a speed of 7–10 knots occupies 29.7%, wind with a speed of 11–16 knots occupies 8.7%, and wind with a higher speed of 17 knots occupies 1.4%.

3.2.1.3 Tides

Tides are an insignificant factor for coastline changes and dynamics along the NDC because the study area is located within the semidiurnal microtidal zone, which does not exceed 0.5 m high (UNESCO/UNDP 1978). Gaweesh (2003) indicated that the highest registered tide level was 0.94 m on Ras al-Bar, while the lowest tide was 0.54 m on the Burullus coast (Fig. 3.1).

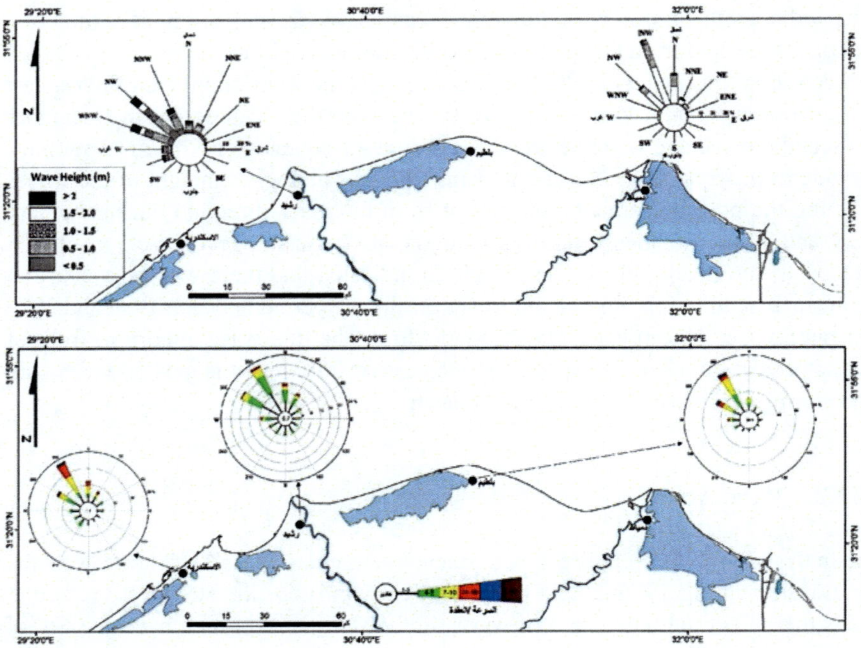

Fig. 3.1 Wave and wind rises along the Nile Delta coast

3.2.1.4 Littoral Currents

Studies of coastline positions and sediment budgets along the NDC indicated that the coastal zone can be divided into a series of discrete sedimentation compartments (Frihy and Khafagy 1991; Frihy and Lotfy 1997). Frihy et al. (2003) identified five different littoral subcells for sediment transportation along the NDC (Abu Qir Bay, Rosetta promontory, Al-Burullus Headland, Damietta promontory, and Port Said). These subcells are considered part of the regional coastal cells extending from Alexandria to Akko in the northern part of Haifa Bay, Israel (Inman and Jenkins 1984). Each cell includes sources, sediment basins, and their transport channels, as well as a continuous direction of coastal transport and sedimentation. The main sources of sediment in each cell are the degraded Nile promontories, which have contributed a significant volume of sand to the coast (Frihy et al. 2003) (Fig. 3.2).

3.2.2 Human Interventions

The construction of dams on the Nile River in the Aswan and barrages at the top and bottom of the Nile River have contributed to cutting off all the river's load of suspended sediments that fed the coast. Moreover, the decrease in water

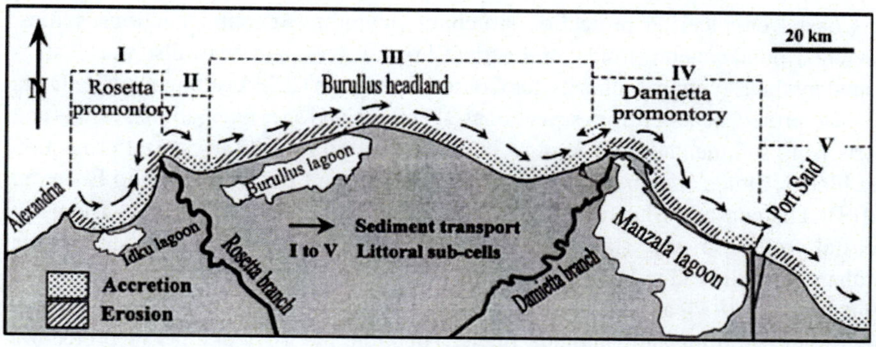

Fig. 3.2 The position of the four littoral subcells along the delta identified by Frihy et al. (2003)

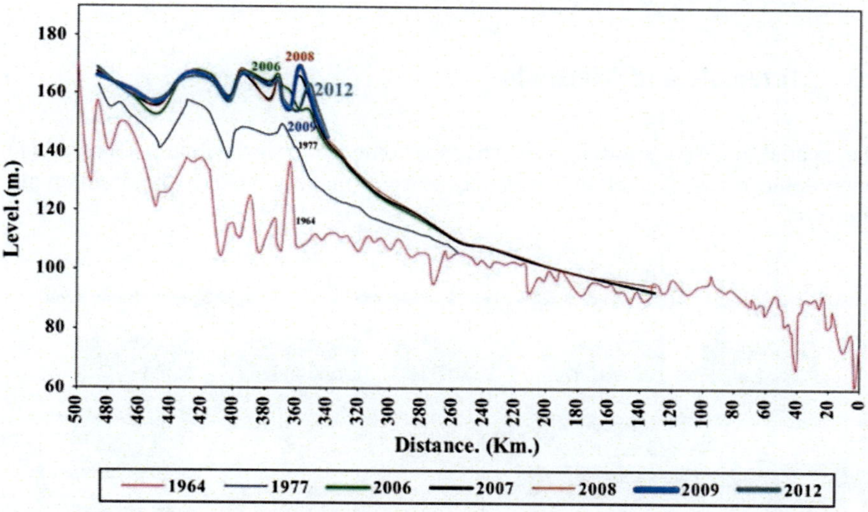

Fig. 3.3 Longitudinal section of Lake Nasser reservoir (1964–2012) by El-Manadely et al. (2017)

supply resulting from the low amount of rainfall in the river basin in the Ethiopian Plateau region and Central Africa also contributed to the decrease in sedimentary supply (Frihy and Khafagy 1991). During the period 1825–1902, the average load of suspended sediment at the Low Aswan Dam reservoir was estimated at approximately 200 million tons/year with a water quantity of approximately 110 billion m^3/year, while in the period 1902–1963, the average load decreased by 20%, reaching approximately 160 million tons/year, and the amount of water decreased to 85 billion m^3/year. During the period 1964–2000, the average load decreased to 124 million tons/year because of the decrease in water, which amounted to 55.5 billion m^3/year. (Frihy and Lawrence 2004) (Fig. 3.3).

Coastal engineering protection structures, including recommendations, jetties, detached breakwaters, groins, submerged breakwaters, and seawalls, were established by the Egyptian Authority for Coastal Protection (ESPA) along the Nile Delta coast to protect beaches from severe coastal erosion and safe navigational canals and ports from sedimentation. The most important materials used in coastal protection are blocks, Dolos (Concrete, Limestone fragments, and Basalt (Frihy and El Sayed 2013). The coastal protection of NDC was performed in several stages. The early coastal protection stage started along the low beaches of Abu Qir Bay, such as the Muhammad Ali Wall in 1830 from storms and waves, the protection of the beaches of Alexandria in 1934, and the construction of jetties at the entrance to the Suez Canal, Idku Inlet, Burullus and Damietta Inlets. Then, the intensive and heavy protection phase of the construction of Seawalls, Detached Breakwaters, and Groins vertical barriers began in 1980 to present.

3.3 Materials and Methods

The spatial datasets used in this chapter come from three main sources: (1) topographic maps, (2) satellite remotely sensed imagery, and (3) field surveying (Fig. 3.4).

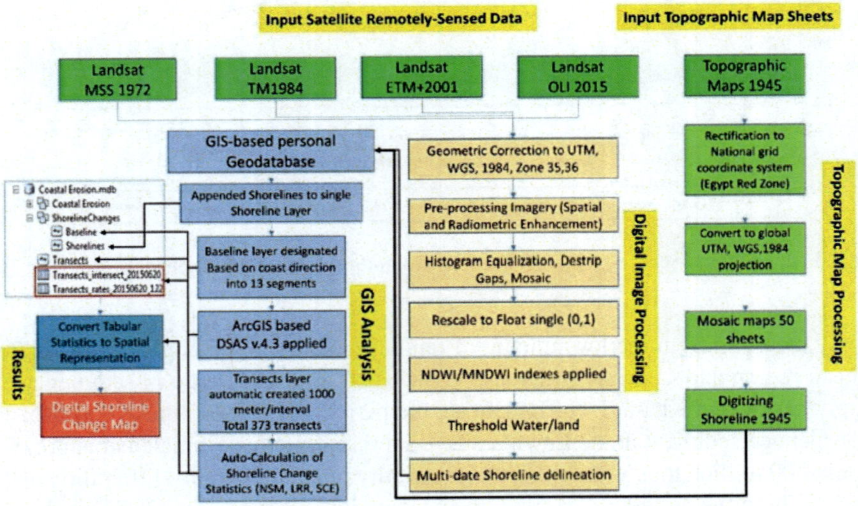

Fig. 3.4 Flowchart of coastline dynamic assessment

3.3.1 Topographic Maps

Historical topographic map scale 1:25,000 (47 sheets) produced by the Egyptian Surveying Authority and completely surveyed in 1945, with a vertical contour interval of 2 m and a horizontal interval of 250 m, was used to map the historical coastline position for the NDC. These regional maps were georeferenced to the Egyptian National Grid Projection System (Egyptian Transverse Mercator, Egypt Red Belt), transformed to the international projection (UTM, WGS, zones 36 N), and then the shoreline was manually digitized.

3.3.2 Pre-Processing of Satellite Imagery

The importance of satellite imagery for coastline detection and dynamic analysis has increased in recent years. Remote sensing satellites provide digital multispectral and hyperspectral imagery that helps to distinguish land and water features (Louati et al. 2014).

In this study, the multitemporal Landsat imagery used in this study includes (1) Multispectral Scanner (MSS), (2) Thematic Mapper (TM), (3) Enhanced Thematic Mapper (ETM), and (4) Operational Land Imager (OLI).

The preprocessing of the imagery consists of (1) geometric correction and transformation to the projection UTM, WGS 1984, zone 36. (2) Layer stacking of MSS, TM, ETM, and ETM+. (3) Conversion from DN values to surface reflectance. (4) Radiometric, spatial, and spectral enhancements and principal component analysis were used to remove signal noise, haze, and clouds. (5) Histogram equalization, color matching, and resolution merging are applied. (6) Mosaicking the scenes of each year into one image (Table 3.1).

3.3.3 Shoreline Delineation

3.3.3.1 Manual Digitizing of Coastline from Topographic Maps

Mapping historical vector-based coastline for the study area was collected from 1:25.000 scale topographic maps produced by the Egyptian Surveying Authority published in 1945. These maps were adjusted geometrically to the Egyptian National Grid Georeference System (Egypt Red Belt) and then transformed to the international projection UTM to be compatible with Landsat imagery. ArcGIS software was used to mosaic the topo sheets as well as manual digitizing of the coastline. Examples of the topographic sheets are shown in Fig. 3.5.

Table 3.1 Specifications of Landsat MSS, TM, ETM+ data

Satellite	Sensor	Spatial resolution	Spectral resolution	Path/raw	Acquisition date
Landsat	MSS	57	4	P188r39, P189r38, P190r38, P190r39, P191r38, P191r39	1972.10.22, 1973.01.03 1972.08.31 1972.09.19
Landsat	TM	30	7	P176r38, P176r39, P177r38, P177r39, P178r39	1984.09.20 1984.09.11 1984.11.21
Landsat	ETM +	15	9	P176r38, P177r38, P177r39, P178r39	2001.01.30, 2001.05.29 2001.04.11 2001.05.20
Landsat	OLI	15	11	P177r38, P176r38, P177r39, P178r39	2015.04.26 2015.04.19, 2015.04.26 2015.02.28

Source http://earthexplorer.usgs.gov/, http://glovis.usgs.gov/

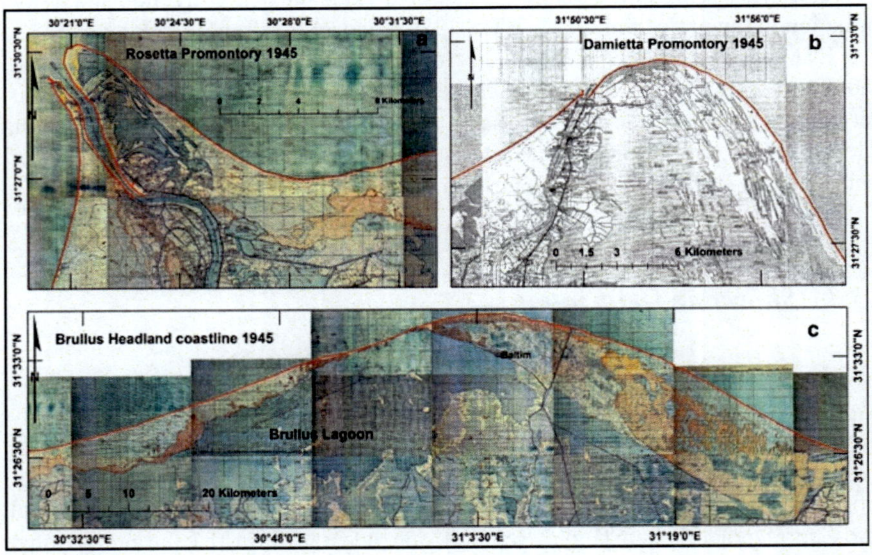

Fig. 3.5 Nile Delta shoreline in 1945. **a** Rosetta promontory, **b** Damietta promontory, **c** Brullus Headland

3.3.3.2 Automatic Extraction of Coastline from Landsat Imagery

Radiometric and spatial enhancements can be applied to satellite imagery to ensure an accurate extraction of the shoreline over time. Radiometric enhancement of satellite imagery was used to remove noise, clouds, and descripting and to rescale the Landsat imagery from a multispectral integer color image to a float binary image. The spectral indices were used globally to separate the land off water bodies. In this study, water indices were applied, including the normalized difference water index (NDWI) and the modified normalized difference water index (MNDWI). The output results of these indices are a single float image with a pixel value range between (−1/+1), a binary threshold was applied to classify water/land, and then the shoreline was identified. The NDWI is represented by Eq. (3.1):

$$\mathrm{NDWI} = \frac{\mathrm{rGreen(band4)} - \mathrm{rNIR(band6)}}{\mathrm{rGreen(band4)} + \mathrm{rNIR(band6)}} \tag{3.1}$$

where

ρ Green = reflected radiation between 0.60 and 0.70 μm.
ρ NIR = reflected radiation between 0.70 and 0.90 μm.

The NDWI was applied only to the Landsat MSS images, as shown in Fig. 3.6.

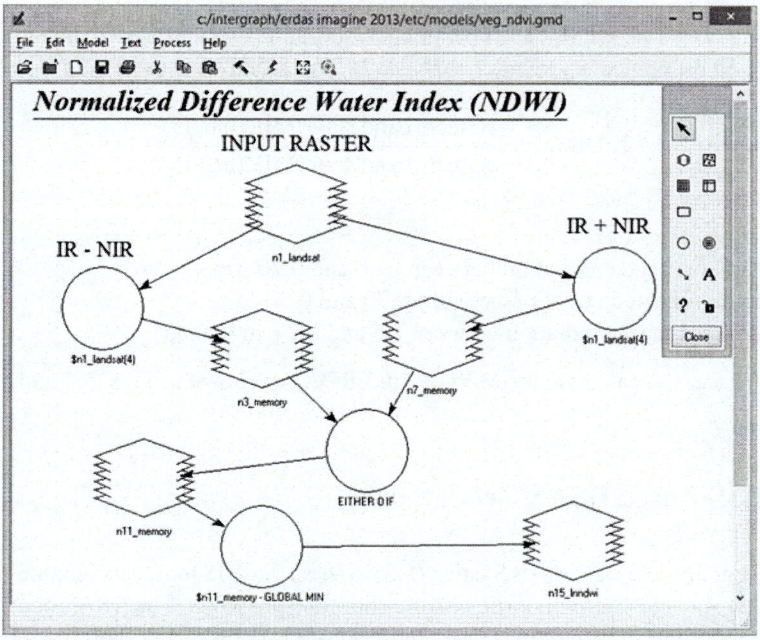

Fig. 3.6 NDWI model calculation using ERDAS imagine v.2014

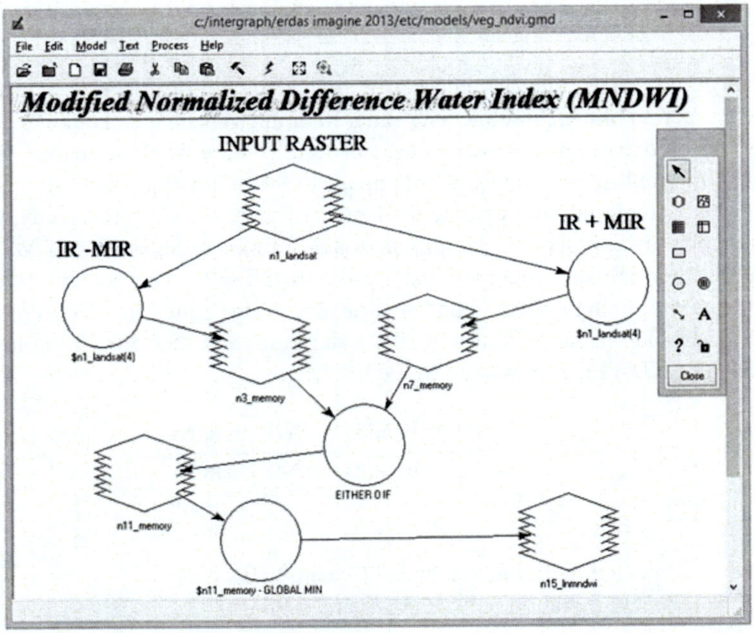

Fig. 3.7 MNDWI model calculation using ERDAS imagine v.2014

For Landsat TM, ETM+, and OLI images, shoreline extraction used the modified normalized difference water index (MNDWI). The MNDWI is expressed as:

$$\text{MNDWI} = \frac{\text{rGreen(band2)} - \text{rMIR(band5)}}{\text{rGreen(band2)} + \text{rMIR(band5)}} \tag{3.2}$$

where

rGreen = reflected radiation between 0.60 and 0.70 μm.
ρ NIR = reflected radiation between 0.70 and 0.90 μm.
ρ MIR = emissive radiation between 1.5 and 1.7 μm (band5).

The process to calculate the MNDWI in ERDAS is shown in Figs. 3.7 and 3.8.

3.4 GIS-Based DSAS Geodatabase

The Digital Shoreline Analysis System (DSAS) is an ArcGIS tool that computes rate-of-change statistics from multiple historic shoreline positions. The DSAS computes the net shoreline movement (NSM), end point rate (EPR), and shoreline change envelope (SCE). The standard error, correlation coefficient, and confidence interval are also computed for the simple and weighted linear regression methods. The results

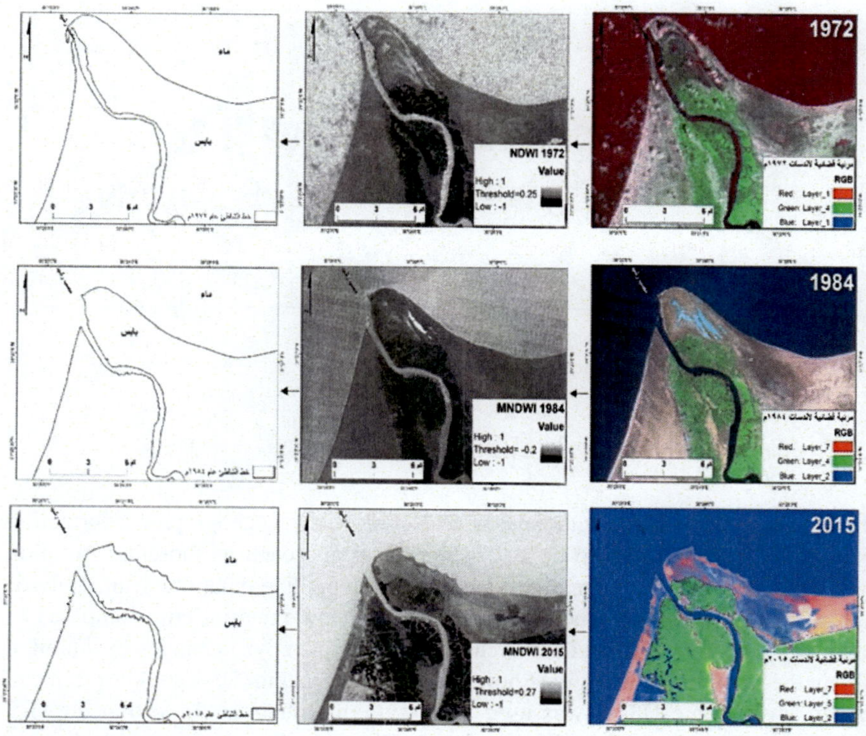

Fig. 3.8 Application of water spectral indices (NDWI/MNDWI) for coastline extraction

of all rate calculations are output to a table that can be linked to the transect file by a common attribute field.

Building a spatial database for the digital analysis system of shoreline changes requires a number of layers collected in a separate and system-linked geodatabase (Thieler et al. 2009):

- ***Append Shorelines (Polyline layer)***: All coastlines are combined into a single layer containing two fields (Uncertainty = double, Date = Text).
- ***Baseline (Polyline layer)***: This is drawn by the user at a distance from the coast and at an angle following the same angle as the coast to ensure the perpendicularity of the intersection lines on the shorelines and contains a field called (Direction = Offshore and Onshore).
- ***Transects (Polyline Layer)***: The program based on the previous layers automatically fulfills them, and the distance between each two lines is known as the horizontal interval. In this study, a horizontal interval of 1 km was chosen. The accuracy of analysis in the program depends on the required threshold and confidence interval. Figure 3.9 shows the sectors used to measure the change in the shoreline in the study area.

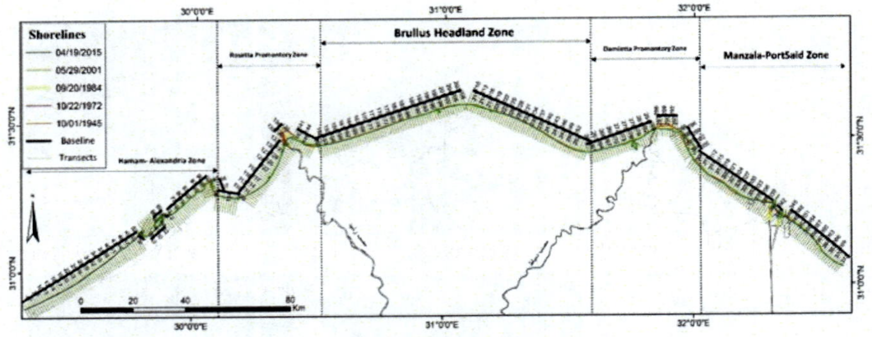

Fig. 3.9 GIS-based DSAS technique for study area

3.4.1 Coastline Dynamic Computation

The integration of modern remote sensing techniques with geographic information systems (GIS) has proven to be a very important approach to shoreline variation studies to cover the study area with modern and repetitive data, the availability of high-resolution data, multispectral databases, and low material cost compared to traditional techniques (Chand and Acharya 2010). To analyze changes in shoreline lines in the study area, the Digital Shoreline Analysis System (DSAS v.4.3) produced and updated by the USGS was used. This system, which works within the GIS environment ESRI ArcGIS® v.10.4.1, aims to automatically calculate beach changes in a set of variables integrated within the system, the most important of which are net shoreline movement (NSM), shoreline change envelope (SCE), End Point Rate (EPR), and linear regression rate (LRR). Many recent studies around the world concerned with studying the change in beach lines have relied on the use of the DSAS automated analysis system, the most important of which are Mujabar and Chandrasekar (2013), Deepika et al. (2014), Louati et al. (2014), Mahapatra et al. (2014), Wang et al. (2014), Kankara et al. (2015), David et al. (2016), Nandi et al. (2016), Darwish et al. (2017).

3.5 Results and Discussion

The shoreline of the Nile Delta changed considerably between 1945 and 2015. The change was more pronounced in some areas than others and during certain periods of time. The output of the DSAS dataset was analyzed spatially along the coast, including (1) net shoreline movement (NSM), representing the total distance between the oldest and most recent shorelines. NSM was calculated for the entire Nile Delta coast, (2) the end point rate (EPR), which was calculated by dividing the distance of shoreline movement by the time elapsed between the oldest and the most recent shoreline, and (3) the shoreline change envelope (SCE). The SCE computes a distance, not a rate. The SCE is the distance between the shoreline farther from and closer to the baseline at each transect. This represents the total change in shoreline movement for all available shoreline positions and is not related to their dates. Shoreline advance and retreat trends were analyzed over four time periods. The analysis assessed changes in shoreline advance and retreat rates before the construction of the Aswan High Dam began in 1964 and after its completion in 1970. Additionally, we analyzed shoreline advance and retreat before and after the construction of sea walls along the Rosetta and Damietta promontories between 1984 and 2001.

A comparative spatial analysis was performed along the coast to analyze and map the areas of advance and retreat of the shoreline. The Nile Delta coast was divided into five geomorphic zones that had relatively homogeneous change patterns. They were (1) Alexandria to Al-Hamam, (2) Abu Quir to Rosetta promontory, (3) Burullus Headland, (4) Damietta promontory and, (5) Al Manzala coastal barrier to Port Said.

Using DSAS, 373 transects were superimposed on the Delta shoreline. Each transect was separated by 1 km. The baseline, which was used to analyze the Nile Delta shoreline, was divided into 19 segments; these segments were divided and designated according to the direction angle of the coastline. The total distance of changed shoreline, either by retreat or advance, along the Nile Delta coast between 1945 and 2015 ranged between advancing 5329 m and retreating 2771 m.

3.5.1 Overall Assessment of Coastline Dynamics (1945–2015)

The analysis of the shoreline change envelope (SCE) along the NDC during the period (1945–2015) shows that stable beaches, which range from 0–200 m, covered 190 km out of a total of 373 km, with 50.94% of the NDC. It was distributed along the rocky Hammam-Alexandria coast, Abu Qir Bay, and the lagoon barriers. The lower coastline change ranges from >200 to 500 m, occupies 125 km, or 33.5% of the study area, and extends over the barriers of Port Said, Gamasa, Alexandria, and Baltim Lakes. Moderate Change: It ranges between (>500 and 2000 m) and occupies 45 km, or 12.06% of the study area, and extends on the sides of the Nile River mouths, ports, and the Manzala Barrier. High and very high zones: Higher Change: This zone represents the changes that exceed 2000 m and occupies 13 km along the coast at a rate of 3.49% of the study area. It is spread along Abu Qir port, Rashid estuary, and the eastern side of the Damietta estuary. This is as shown in Table 3.2 and Fig. 3.10.

The analysis of the statistical characteristics of coastline advance/retreat rates during the period from 1945 to 2015 shows that there is an equilibrium in the erosion and accretion zones along the NDC:

Table 3.2 Coastline change detection between 1945 and 2015

Class	Shoreline change	Range of change (Meter)	Number of transects (out of 373)	Distance along the coast/km	%
1	No change (stable)	<200	190	190	50.94
2	Low change	>200–500	125!	125	33.51
3	Moderate change	>500–2000	45	45	12.06
4	High change	>2000–4000	11	11	2.95
5	Very high change	>4000	2	2	0.54

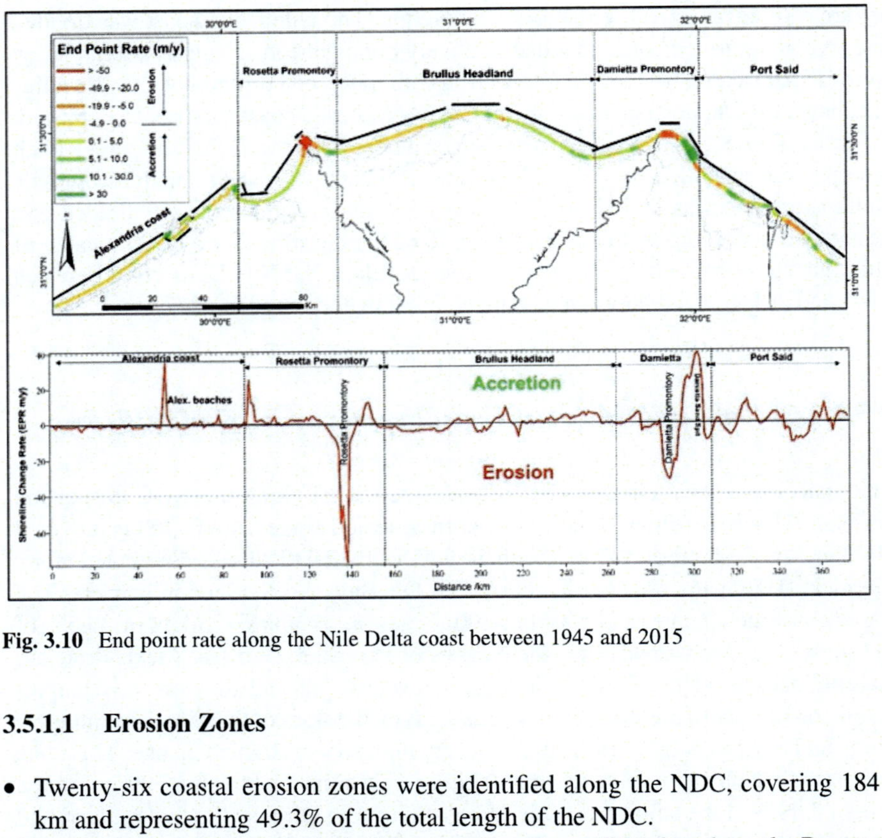

Fig. 3.10 End point rate along the Nile Delta coast between 1945 and 2015

3.5.1.1 Erosion Zones

- Twenty-six coastal erosion zones were identified along the NDC, covering 184 km and representing 49.3% of the total length of the NDC.
- The maximum retreat distance of the coastline was −5328.6 m along the Rosetta promontory, while the average retreat distance was −324.5 m.
- The highest annual erosion rate was −76.6 m/year in the Rosetta estuary, with an average annual rate of −4.8 m/year.
- The lowest erosion rates ranged between 0 and 2 m/year cover a total length of 97 km, with 52.7% of the coastline eroded. These areas were distributed along Hammam-Alexandria, the eastern side of the Rosetta promontory (Abu

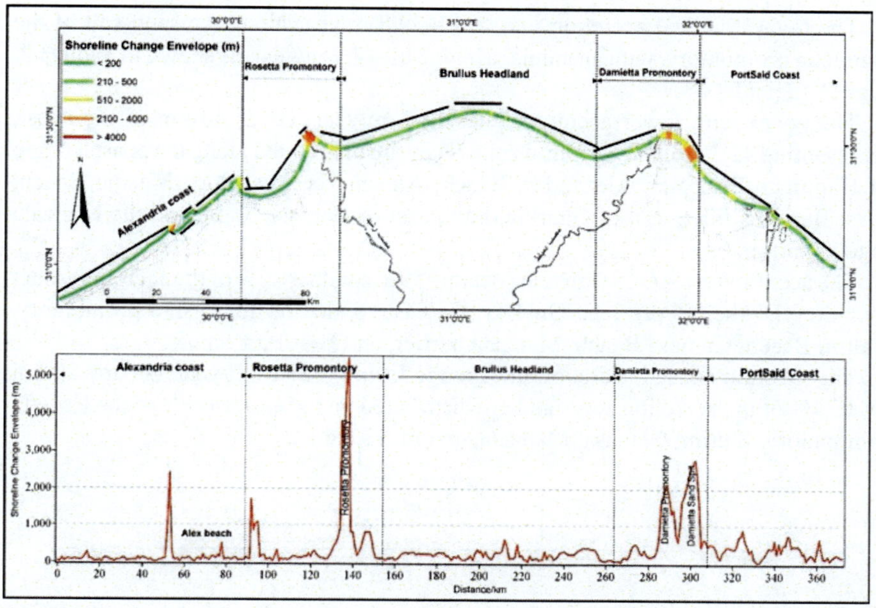

Fig. 3.11 Shoreline change envelope (SCE) along the Nile Delta coast from 1945 to 2015

Khashaba), the Burullus headland, the eastern side of Gamasa Bay, Ras al-Bar beach, and some short ranges on the Manzala-Port Said strip.

- Erosion rates between >2 and 5 m/year occupy 52 km, with 28.3% of the coastline eroded, and are distributed along the western side of the Rashid Estuary, Burullus Barrier, Burullus-Baltim Beach, New Damietta Beach, and Manzala Barrier.
- The medium coastal erosion rates (>5 m/year) occupy 21 km by 11.4% of the eroded coastline and are distributed along the sides of the Rosetta promontory, Baltim, Damietta promontory, Manzala coastal lagoon barrier, and eastern Port Said.
- Severe coastal erosion rates occupy 14 km, with 7.7% of the coastline eroded, and are distributed along the Rosetta and Damietta promontories, as shown in Table 3.2, Figs. 3.10, and 3.11.

3.5.1.2 Accretion Zones

The analysis of coastal dynamics along the NDC shows that 25 accretion zones cover 189 km of the NDC, representing 50.7% of the coast, which is therefore slightly longer than the distance exposed to erosion.

The maximum advanced distance of the coast was +2770.9 m along the eastern side of the Damietta estuary, while the average distance provided by the coastline was +319.2 m during (70 years).

The highest annual accretion rate was +39.8 m/year along the sand spit of the Damietta promontory, with an annual average of −4.7 m/year, as shown in Table 3.2, Figs. 3.12, and 3.13.

The zones with lower accretion rates (0–2 m/year) cover 80 km of the coast, representing 42.3% of the accretion zone. They are distributed along the coastal range of Hammam-Dekheila, Alexandria Beach, Alexandria West Port, Maadia Beach, West Burullus Inlet, east of Gharbia drain, east of Gamasa drain, and the Manzala lagoon barrier.

The accretion rates (>2–5 m/year) occupy 59 km with 31.2% of the accretion zone, including Dekheila Port, Abu Qir Bay, the eastern side of the Rashid promontory, Baltim Beach, Gamasa Beach, Manzala barrier, and East Port Said.

The medium coastal accretion rates range from >5 to 10 m/year, occupy 29 km by 15.4% of the accretion zone, and are distributed on the eastern side of the Rosetta promontory, Baltim, Gamasa, Manzala, and Port Said.

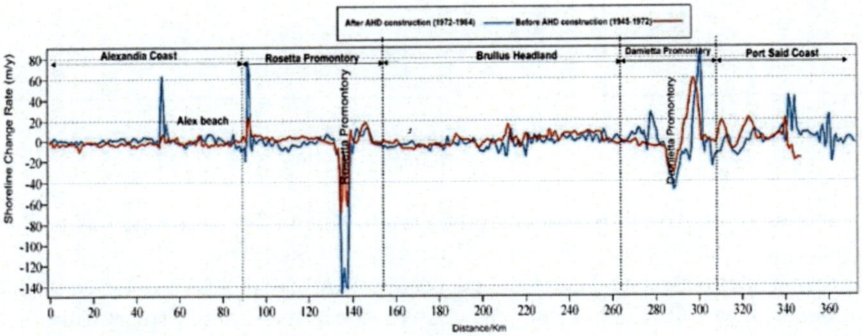

Fig. 3.12 Impact of the Aswan high dam on coastline dynamics

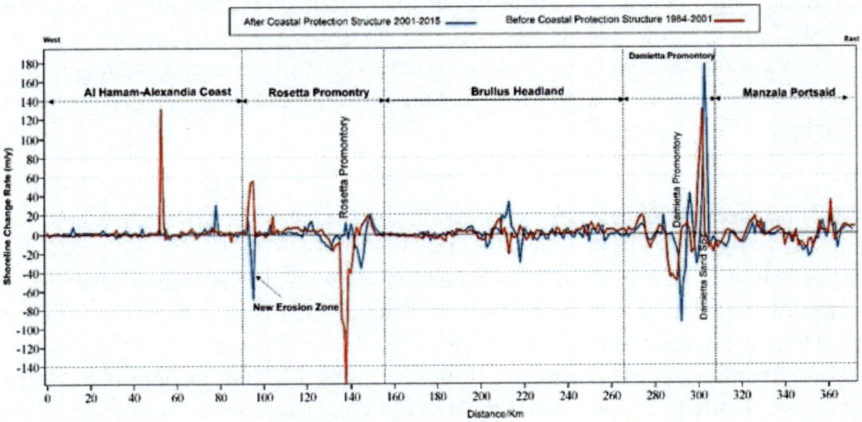

Fig. 3.13 Impact of coastal protection structure on coastline dynamics

The severe accretion rates occupied 21 km, with 11.2% of the accretion zone. They are distributed along Dekheila Port, Abu Qir Port, east of the Rosetta promontory, and the Damietta sand spit, as shown in Table 3.2, Figs. 3.12, and 3.13.

3.5.2 Zonal Coastline Change Analysis

3.5.2.1 Alexandria to Al-Hamam Coastal Zone

This coastal zone extends from the sand beaches of Al Hammam city in the west to the Abu Qir headland on the eastern side of Alexandria city, and most of Alexandria's beaches are relatively stable because the coastline is formed by hard rocks belonging to the coastal limestone scarp (Frihy et al. 2010). To analyze the coastline changes in this coastal zone, 92 perpendicular transects were used to measure coastline advance/ retreat with a horizontal interval of 1 km. This zone was divided into three main sectors according to the littoral cells:

Hammam to Al-Dekheila (52 Transect)

This sector belongs to the Alamien subcell for sediment transfer along the northwest coast toward the east (Frihy et al. 2010), and 46 vertical sectors out of 52 sectors indicate that the coastline retreated during 1945–2015. It is clear from the analysis of erosion and accretion rates in the study area that the highest erosion rate was (−3.4 m/year) with an average erosion rate (−0.96 m/year), and (6) sectors indicated an advance in coastline. The highest sedimentation rate was (+2.2 m/year) with an average of (+0.24 m/year), which is one of the very low rates, as shown in Fig. 3.3.

Spatiotemporal analysis of erosion and accretion rates along this coastal strip shows coastline retreat at an average rate of −2.8 m/year from 1945 to 1972, it decreased gradually to −1.2 m/year from 1972 to 1984, then decreased to (−0.93 m/year) from 1984 to 2001, and (−0.97 m/year) during the period from 2001 to 2015, which indicates that this coastal strip is almost stable. This coastal stripline consists of limestone rocks, and there is no clear relationship between the impact of the construction of the High Dam and coastal engineering protection works on coastline morphodynamics. It is also clear from the analysis that the coastline in this sector did not experience significant accretion during the period from 1945 to 1972, while in the period from 1972 to 1984, the average accretion rate was (+1.84 m/year) and then decreased to (+0.82 m/year) during 1984–2001. During the period from 2001 to 2015, the accretion rate was +1.4 m/year, all of which were considered very low sedimentation rates (0–2 m/year).

Artificial Port of Dekheila (4 Transect)

Dekheila Port was built artificially along the Dekheila headland as a natural extension of the western port of Alexandria in 1986. The analysis of the annual coastal change rates during the time period between 1945 and 2015 indicated that the coast advanced toward the sea at a distance of +2431 m at the highest accretion rate of +34 m/year with an average annual advance accretion rate of (+13.7 m/year). These changes have been made as a result of the artificial construction of the port. It is also clear from the analysis of erosion rates that this sector has not experienced a retreat during this time period.

The spatiotemporal coastline dynamics of erosion and accretion rates along this coastal sector show an advance in the coastline with an average annual accretion rate of (+4.6 m/year) from 1945 to 1972, then the average accretion rate increased during 1972–1984 to (+23.1 m/year) as a result of the start of the port construction work, it increased to (+70.8 m/year) from 1984 to 2001 as a result of the completion of the construction of Dekheila port, then the annual average decreased to (+4.55 m/year) during 2001–2015 after the completion of port construction. The analysis of erosion rates in this coastal sector clearly showed that the average annual erosion rate reached −1.7 m/year from 1945 to 1972, while in the period from 1984 to 2001, the average rate was −2.63 m/year decreased to (−0.6 m/year) between 2001 and 2015. All erosion rates fall within very low and low rates, and there is no clear impact of the construction of the High Dam on the change in erosion rates in this sector.

Alexandria Beach (36 Transect)

Alexandria coast extends from the eastern side of Dekheila Port, including (Alexandria West and East Ports, Montazah and Maamoura beaches) to Abu Qir headland in the east. The beaches of Alexandria have been protected from coastal erosion since 1934 with a vertical wall of 20 km and are known as the Corniche highway (Frihy et al. 2010). Concrete blocks have also been installed along the coast of Alexandria since 1984, and breakwaters have been used to protect Dekheila Port, West Port, Eastern Port, and Abu Qir Port. In the period 2005–2007, submerged breakwaters were established with the aim of reducing erosion and wave action on the beaches of Asafra, Mandara, and the park (Frihy et al. 2010). Two 65-m-long Jetties at the mouth of the Nubaria Bank were built in 1986 approximately 20 km west of Alexandria (Fanos et al. 1995). Coastline dynamic analysis along Alexandria beaches shows that the coastline retreated along 18 sectors by a distance of 124.5 m during 1945–2015, with the highest erosion rate (−1.79 m/year) with an average annual erosion of (−0.86 m/year). On the other hand, the analysis of accretion rates in this sector indicates that the maximum advance of the coastline seaward along 15 sectors amounted to 389.5 m, and the highest annual accretion rate was (+5.6 m/year) with an average annual accretion rate (+1.2 m/year) as a result of the establishment of many artificial projects to feed the beach, such as Artificial Beach Nourishment in the areas of Shatby, Stanley, Asafra, Mandara, and Abu Qir, by bringing a quantity of sand estimated at (0.31 m^3 × 106 m^3) from the desert at a distance of 150 km southwest of Alexandria. Some of these projects were associated with the installation of vertical barriers and short Groins. In general, renourishment projects have been effective in protecting beaches from coastal erosion (Frihy and Dean 1992).

Spatiotemporal coastline dynamic changes along this coastal sector show a retreat in coastline with an annual average erosion rate (−2.6 m/year) from 1945 to 1972, then the average annual rate increased to (−3.7 m/year) from 1972 to 1984, which indicates that the coast of Alexandria was affected by the construction of the High Dam in 1964, which contributed to an increase in the rate of coastal erosion as a result of the lack of flood supply to the coast, and after the reconstruction of the "Corniche" in 1984, the rate of coastline retreat decreased to (−1.9 m/year) during 1984–2001, then the retreat rate continued to decrease to −1.49 m/year from 2001 to 2015, and this noticeable decrease in erosion rates was due to coastal protection. On the other hand, the coastline in the Alexandria beach sector advanced with an average annual advance rate (+2.3 m/year) during the period from 1945 to 1972 and then increased to +2.96 m/year from 1972 to 1984, it decreased to (+2.5 m/year) during 1984–2001 and then increased again to (+3.3 m/year) from 2001 to 2015, which occurred as a result of the increase in beach feeding projects with materials brought from the desert to protect them from erosion as shown in Table 3.3.

3.5.2.2 Abu Quir Bay to Rosetta Promontory

This coastal zone extends along the coast of Abu Qir Bay on the eastern side of Alexandria city; it starts from the Abu Qir headland in the west to the eastern side of Rosetta promontory near Abu Khashaba beach, including Abu Qir and the Rosetta littoral cells. This coastal zone was divided into three subsegments.

AbuQir Port

The artificial construction of the Abu Qir port in the Abu Qir headland began in 1983, and 6 perpendicular transects were used to assess the coastline rate of change from 1945 to 2015. The analysis of the erosion and accretion rates along this segment shows a seaward advance of the coastline (+1741 m), and the maximum accretion rate was (+25 m/year) with an average annual accretion of (+9.23 m/year). The annual accretion rate increased due to human intervention in the construction of the seaport in the Abu Qir headland.

Spatiotemporal changes in coastline erosion and accretion rates along this coastal sector indicated that the coastline advanced seaward with an annual accretion rate of +9.7 m/year between 1945 and 1972, and the annual accretion rate increased to + 23.4 m/year during 1972 and 1984 because of port construction during this period. After that, the annual accretion rate increased to (+34.4 m/year) between 1984 and 2001. The average annual accretion rate decreased to (+3.3 m/year) after the completion of port construction during 2001–2015. The coastline change analysis

Table 3.3 Coastline change rates along the NDC between 1945 and 2015

Time period	1945–1972 Before AHD				1972–1984 After AHD				1984–2001 Before Seawall				2001 to 2015 After Seawall			
Coastal segment	Erosion		Accretion		Erosion		Accretion		Erosion		Accretion		Erosion		Accretion	
	A	H	A	H	A	H	A	H	A	H	A	H	A	H	A	H
Hamam to Dekheila coast	2.8	9.6	0	0	1.2	2.8	1.8	4.4	0.93	2.4	0.82	4.9	0.97	3.94	1.4	7.8
Dekheila headland	1.7	1.7	4.6	5.3	0	0	23.1	63.8	2.6	4.7	70.8	131.75	0.6	0.63	4.6	6.93
Alexandria city beaches	2.6	5.2	2.3	6.5	3.7	10.2	2.96	7.06	1.9	7.5	2.5	7.4	1.49	4.4	3.3	29.3
AbuQuir headland	0	0	9.7	24.5	2.5	3.5	23.4	88.7	2.1	2.4	34.4	55.09	44.9	68.4	3.3	13.2
AbuQuir Bay	3.2	11.2	2.97	5.3	4.4	13.8	2.6	5.9	4.7	17.6	5	18.02	5.3	16.9	3.13	13.8
Rosetta promontory	28.7	72.2	10	18	71.7	140.6	10.1	13.3	52.2	158.98	9.97	18.7	11.5	35.2	10.3	13.2
Brullus headland	3.6	7.7	4.9	17.3	4.5	16.8	3.7	11.98	3.1	20.7	3.8	15.7	3.6	30.4	4.99	32.8
Damietta promontory	7.6	13.2	20	60.9	13.13	45.2	19.6	78.03	20.6	50.2	15.4	129.48	12	93.9	27.6	177.6
Port said beaches	6.9	18.8	9.5	23.4	7.9	24.2	8.3	43.8	8.2	19.17	6.4	33.7	8	24.97	5.5	17.54
Overall	5	72.2	7.2	60.9	8.3	140.6	7.5	88.8	8.23	158.98	6.9	131.75	7	93.9	5.97	177.6

AHD = Aswan High Dam, **A** = Average Rate (m/y), **H** = Highest Rate (m/y) along this coastal sector indicated that there was no significant retreat during the period from 1945 to 1972, while the coastal erosion rate was (−2.45 m/year) during 1972–1984 and decreased to (−2.1 m/year) from 1984 to 2001; however, during the period 2001–2015, the coastline retreated at an annual average rate (−44.9 m/year) because of port maintenance and expansion landward.

Maadia-Idku Segment

This coastal segment was one of the first coastal areas along the NDC to be protected due to coastal erosion and sea level rise since 1830. A stone barrier called the Muhammad Ali barrier was constructed with a length of 10 km to save the Lake of AbuQir, and in 1980, this barrier turned into a seawall with an approximate height of 3.5 m above sea level to face wave action and potential sea level rise (Frihy 2003). Moreover, short jetties were built in 1962 on both sides of the entrance of Lake Idku, and they were rebuilt in 1980 to protect the lake's navigational canals from sedimentation processes (Fanos et al. 1995).

Thirty-six perpendicular transects were used to measure the annual rate of change in the coastline along this coastal segment. The analysis shows that the maximum retreat of the coastline from 1945 to 2015 was −674 m and amounted to the highest rate of coastal erosion in this sector (−9.6 m/year) with an average annual erosion of −4.5 m/year, which was classified as a low erosion rate. A number of (29) sectors indicated that the maximum advanced distance of the coastline seaward was (+332 m), and the highest annual accretion rate was (+4.78 m/year) with an average annual accretion of (+2 m/year), which is classified among the low rates.

Spatiotemporal coastline dynamics along this segment indicated that the coastline was affected by the construction of the Awan High Dam, which worked to prevent the flow of sediment to the coast. The average rate of coastline retreat in this sector was (−3.17 m/year) from 1945 to 1972, and then the erosion rate gradually increased to (−4.4 m/year) from 1972 to 1984, increased to (−4.7 m/year) from 1984 to 2001, and increased to (−5.3 m/year during the period 2001–2015. The average annual accretion rate was +2.97 m/year Between 1945 and 1972, it decreased to (+2.6 m/year) from 1972 to 1984, then the average rate of progress increased to (+5 m/year) between 1984 and 2001, and then the average rate decreased to (+3.13 m/year) in the period from 2001 to 2015.

Rosetta Promontory Segment

The Rosetta promontory sector is one of the most active coastal dynamic zones along the NDC, and it is clear from the analysis of erosion and sedimentation rates that the highest rate of erosion is concentrated in this sector. Fifteen perpendicular transects were used to calculate the rate of change along the promontory and its eastern and western sides.

Eight transects indicated that the maximum distance of coastline retreat along the promontory tip was −5329 m during the period from 1945 to 2015, and the highest annual erosion rate was −76.6 m/year with an average annual erosion rate of −16.5 m/year.

Seven transects indicate that the maximum advanced distance of the coastline was 828.2 m along the eastern side of the estuary at (Abu Khashaba), while the highest sedimentation rate was (11.91 m/year) with an average of 7 m/year.

The outer margins of the Rosetta promontory were protected from severe erosion and marine processes with two seawalls parallel to the coast and two artificial embankment bridges covered with concrete blocks weighing between 4 and 7 tons. They were built between 1988 and 1991 on the eastern side of the estuary with a length of 3.35 km and the western side of the estuary with a length of 1.5 km, a height of 6.75 m, and an average width ranging between 48 and 70 m. Five vertical groins were installed in 2003 along the eastern margin of the eastern seawall, and their lengths range between 400 and 500 m with a separation distance of 800–900 m, while along the western side of the Rosetta promontory, ten vertical short groins were built with lengths ranging from 80 to 150 m in 2005 (Frihy et al. 2010). Spatiotemporal coastline dynamics along the Rosetta promontory show the highest coastal erosion rate (−72.2 m/year) before the construction of the High Dam, with an average of −28.7 m/year during the period 1945–1972. During the period 1972–1984, after the construction of the Aswan High Dam, the highest coastal erosion rate increased to −140.6 m/year with an average annual rate of (−71.7 m/year). The highest coastal erosion rate increased to −158.9 m/year in the period from 1984 to 2001, and the average annual rate was (−52.2 m/year). The average coastal erosion rate dramatically decreased by −2.7 m/year after the construction of the seawalls along the Rosetta promontory during 2001–2015. On the other hand, new coastal erosion zones appeared on both sides of the concrete walls. The highest erosion rate was −35.2 m/year, with an average of −11.49 m/year along the eastern side of the promontory near Abu Khashaba beach.

An explanation of the severe coastal erosion along the Rosetta Estuary is related to the wave and current action in sand transportation away from the mouth of the Rosetta branch, which stopped supplying new suspended sand deposits after the construction of the Aswan High Dam. The eroded amounts of sand move from the front of the promontory toward the west coast of Abu Qir Bay and toward the east of the eastern flank side along Abu Khashaba beach, which is called the divergence nodal. Dewidar and Frihy (2010) identified two sediment nodes (convergence nodal) at two equal distances from the promontory (5 km east and west of the promontory, as shown in Fig. 3.14).

3.5.2.3 Burullus Headland Coastal Zone

The Burullus headland coastal zone is located in the central area of the NDC; it extends from the eastern side of the Rosetta promontory to the end of the Gulf of Gamasa. A total of 109 perpendicular transects were used to measure and analyze

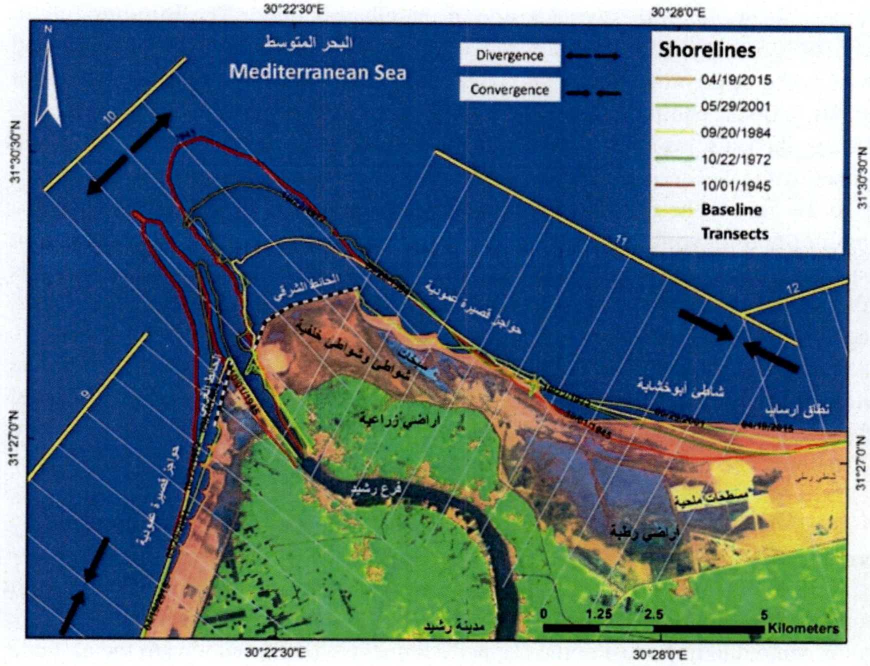

Fig. 3.14 Shoreline positions along Rosetta promontory between 1945 and 2015

changes in the coastline. It is clear from the analysis that the maximum distance of coastline retreat during the period 1945–2015 was −500 m, and the highest rate of erosion was −7.2 m/year, with an average annual rate of −2.7 m/year. The erosion areas are concentrated on the Burullus Barrier, Burullus Beach (Al-Burj Village), and on both sides of the Western Kitchener Drain.

The analysis of the accretion rates along this coastal zone shows that the maximum distance advanced by the coastline was +639.4 m during the period 1945–2015. The highest annual accretion rate was (+9.19 m/year), with an average annual accretion rate of + 2.95 m/year. The accretion areas are distributed along the western side of the Burullus Inlet and Gamasa Embayment. The spatiotemporal coastal dynamics along this coastal zone show the highest rate of coastal erosion (−7.7 m/year) with an average of (−3.6 m/year) during the period 1945–1972 before the construction of the High Dam. The highest rate of coastal erosion increased to (−16.8 m/year) with an average of −4.5 m/year during the period after the construction of the High Dam (1972–1984), which indicates that this range was affected by the construction of the High Dam in 1964. With the construction of several coastal protection engineering means along the Burullus-Baltim coastal zone, the average annual average erosion rate has decreased to −3.1 m/year during the period from 1984 to 2001. The average coastal erosion rate increased to −3.6 m/year during the period from 2001 to 2015 because of the appearance of new erosion zones on the margins of the protected sector of seawalls on the western side of the Gharbia drain. The Burullus-Baltim segment is one of the most dynamic coastal zones along the NDC. It was protected by several coastal protection means; three long jetties were built mainly to control the navigational entrance of the Burullus lagoon along both sides of the Burullus port and the Burullus inlet with lengths ranging between 300 and 180 m (Frihy and Deabes 2011).

To the east of the Burullus Lagoon entrance, there is a concrete seawall with a length of 600 m, which was built in 1982 to protect Al-Burj village beaches from coastal erosion, followed by a basaltic riprap seawall up to 1.3 km long, which was built in 1984 to protect the eastern side of the village of Al-Burj from flooding, storm surges, and coastal erosion (Frihy et al. 2010).

Seventeen separate detached breakwaters were constructed in four phases in the period from 1993 to 2007 with submergence at a depth of 3–4 m below sea level (groin numbers 1–15 range in length from 250 to 300 m, grout numbers 16–17 are 500 m long, and barrier number 15 takes the shape of the letter T) (Frihy and Deabes 2011).

These barriers contributed to the protection of the shore with the dune belt by forming a series of tombolo forms and accretion beaches (Frihy et al. 2010).

To protect the eastern side of the detached breakwaters zone, nine vertical short groins with a length of 40 m were installed in 2005 to stabilize the shoreline, and these groins worked to trap most of the trapped sediments that moved toward the east due to the shore current (Frihy and Deabes 2011). The analysis of accretion rates along the Brullus headland coastal zone shows that the average annual rate of accretion was +4.9 m/year between 1945 and 1972. It decreased after the construction of the High Dam to +3.7 m/year between 1972 and 1984 (+3.8 m/year) during the period 1984–2001. The average accretion rate increased to +4.99 m/year between 2001 and 2015.

3.5.2.4 Damietta Promontory

The Damietta promontory is one of the most dynamic coastal zones in the study area, and the analysis of coastal erosion and accretion rates along the Damietta promontory indicated that the maximum retreated distance of the coastline in front of the Damietta promontory was −2170.2 m, and the highest erosion rate was −31.2 m/year with an average of −9.55 m/year during the period from 1945 to 2015.

The analysis of accretion rates along the promontory shows an advance in coastline with a maximum distance seaward of +2771 m, and the highest accretion rate was + 39.8 m/year with an average of +12.8 m/year. There are two accretion zones along the Damietta promontory, one of which is located along the western side of the Damietta Port and the other on the Sand Spit on the eastern side of the promontory. In 1941, the entrance to the navigational channel of the Damietta estuary was protected using two vertical jetties on the mouth of Damietta with lengths of 240 and 290 m for the western and eastern parts, respectively. On the west bank of Ras El Bar resort, there is a combination of engineering structures constructed from 1991 to 2002, including vertical short groins, basalt fractures, and a series of eight separate groins constructed in the area between Ras El Bar and Damietta Port. In addition, a concrete block Dolos Seawall with a length of six kilometers and an east–west direction was built on the eastern side of the Damietta promontory with a load varying from 4 to 7 tons and ends at the Sand Spit, and both the walls of Rashid and Damietta worked effectively to protect the promontories from coastal erosion and shoreline retreat (Frihy and Lawrence 2004).

Spatiotemporal coastline dynamics along the Damietta promontory show a maximum erosion rate of −13.2 m/year during the period 1945–1972, with an average erosion rate of −7.6 m/year. After the construction of the Aswan High Dam, the highest rate of erosion was −45.2 m/year with an average of −13.13 m/year during the period 1972–1984, which is considered one of the higher rates. During the period from 1984 to 2001, the highest rate of erosion was −50.2 m/year, with an average of −20.6 m/year. Although concrete walls succeeded in reducing the coastal erosion rates, new erosion zones appeared on the margins of concrete seawalls, and the highest erosion rate in the new unprotected ranges was −93.9 m/year with an average of −12 m/year during the period from 2001 to 2015.

The analysis of accretion rates shows that the highest sedimentation rate in the period 1945–1972 was +60.9 m/year with an average of +20 m/year. During the period 1972–1984, the highest rate was +78.03 m/year, with an average rate of + 19.6 m/year, because of the construction of the Aswan High Dam. During the period after the construction of the Aswan High Dam from 1984 to 2001, the rate increased to +129.4 m/year with an average accretion rate of 15.4 m/year. The highest rate of accretion was +177.6 m/year with an average rate of +27.6 m/year along the Damietta sand spit during the period from 2001 to 2015, as shown in Fig. 3.15.

3.5.2.5 Manzala Lagoon Barrier to Port Said

The Manzala–Port Said coastal zone extends along the eastern side of the Nile Delta coast, and the analysis of coastline dynamics along this coastal zone shows that the highest erosion rate in the period from 1945 to 2015 was −10.7 m/year with an average erosion rate of −5.5 m/year. The analysis of accretion rates also shows that the maximum advance of the coastline in this range was +906.1 m, and the highest accretion rate was +13 m/year with an average rate of +4.5 m/year.

Spatiotemporal coastline changes along this coastal zone indicated that the average annual erosion rate was −6.9 m/year during 1945–1972; it increased to −7.9 m/year during the period 1972–1984 because of the impact of High Dam construction. Then, the average rate increased to −8.2 m/year in the period 1984–2001 and −8 m/year during the period 2001–2015. To protect this coastal zone from severe coastal erosion, several means of protection were established, such as jetties on both sides of the Suez Canal; the western jetty was established to protect the entrance to the canal from sediments transported by the shore current toward the east with a length of 7.7 km, and it was built in stages starting in 1864 and reached the current length

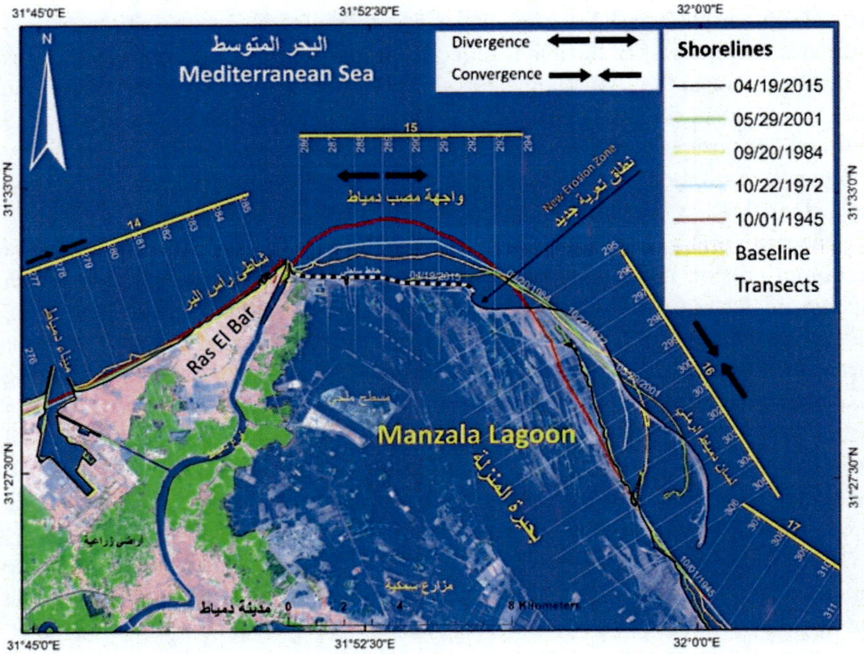

Fig. 3.15 Shoreline positions along the Damietta promontory between 1945 and 2015

in 1924. The eastern jetty was built in 1864 to protect the entrance to the Suez Canal (Frihy et al. 1991).

There are also two short jetties on both sides of the El Gameil Inlet with lengths ranging from 225 to 200 m that are designed to protect the Manzala Lagoon navigational canal from accretion processes. There are six separate detached breakwaters near the village of Paradise on the eastern side of El Gameil Inlet, and on the western side, there are five short vertical grooves to stabilize the navigational corridor (El-Asmar and Hereher 2010), a 4 km long stone seawall to protect the coastal road near El Gameil Airport from the dangers of flooding and waves (UNDP 2009). The analysis of the annual accretion rates along this coastal zone indicated that the average advance rate of the coastline was +9.5 m/year during the period 1945–1972; it gradually decreased to +8.3 m/year between 1972 and 1984, after which it decreased to +6.4 m/year between 1984 and 2001, and in the period following 2001–2015, the average rate decreased to +5.5 m/year.

3.6 Summary

This chapter studied the spatiotemporal coastline dynamics along the NDC during the period between 1945 and 2015 using multitemporal Landsat imagery and advanced geographic information system computation techniques. Five coastline positions were extracted from topographic maps and satellite imagery for use in the analysis of the Nile Delta coast. Coastline change under impact of natural and anthropogenic deriving factors. The NDC were divided into five coastal homogenous zones according to the littoral cells. This study assessed the impact of the Aswan High Dam and coastal engineering protection measures on coastal erosion hazards.

The results of spatiotemporal analysis of the 373 transects used in this study indicated that 184 (49%) refer to erosion, while 189 represent accretion. The maximum distance provided by the coastline was +2770 m, and the highest annual rate was + 39.8 m/year with an average annual advance rate of +4.7 m/year. The construction of the Aswan High Dam affected the acceleration of coastal erosion rates during the period following its construction, as the highest rate of coastline retreat along Rosetta promontory was −72.19 m/year during 1945–1972, while the highest rate was −140.6 m/year after the completion of the construction of the High Dam from 1972 to 1984. It increased to −159 m/year from 1984 to 2001 after the construction of seawalls and other coastal protection means along NDC. Although seawalls contributed to reducing coastline retreat along the coast, this study showed that new erosional zones were formed at the margins of concrete seawalls, and they had a very high coastal erosion rate of −35 m/year during the period from 2001 to 2015.

References

Chand P, Acharya P (2010) Shoreline change and sea level rise along coast of Bhitarkanika wildlife sanctuary, Orissa: an analytical approach of remote sensing and statistical techniques. Int J Geomatics Geosci 1(3):436–455

Darwish K, Smith SE, Torab M, Monsef H, Hussein O (2017) Geomorphological changes along the Nile Delta coastline between 1945 and 2015 detected using satellite remote sensing and GIS

Dewidar Kh, Frihy OE (2010) Automated techniques for quantification of beach change rates using Landsat series along the northeastern Nile Delta, Egypt. J Oceanogr Mar Res 1(2):28–39. https://www.longdom.org/archive/ocn-volume-1-issue-2-year-2010.html

David TI, Mukesh MV, Kumaravel S (2016) Sabeen HM (2016) Long-and short-term variations in shore morphology of Van Island in Gulf of Mannar using remote sensing images and DSAS analysis. Arab J Geosci 9:756

Deepika B, Avinash K, Jayappa KS (2014) Shoreline change rate estimation and its forecast: remote sensing, geographical information systems, and statistics-based approach. Int J Environ Sci Technol 11:395–416

El-Asmar H, Hereher M (2010) Change detection of the coastal zone east of the Nile Delta using remote sensing. Environmental Earth Science 62(4):769–777

El-Manadely MS, Aziz MS, Negm DA (2017) Navigation water-way for Lake Nasser/Nubia on the Nile River. Lakes Reserv Res Manag 22(4):377–389

Fanos AM, Khafagy AA, Dean RG (1995) Protective works on the Nile Delta coast. J Coastal Res 11(2):516–528. JSTOR

Frihy OE, Deabes EA, El Gindy AA (2010) Wave climate and nearshore processes on the Mediterranean coast of Egypt. J Coastal Res 26(1):103–112, January 2010

Frihy OE, El-Sayed MK (2013) Vulnerability risk assessment and adaptation to climate change induced sea level rise along the Mediterranean coast of Egypt. Mitig Adapt Strateg Glob Change 18:1215–1237

Frihy OE, Lotfy MF (1997) Shoreline changes and beach-sand sorting along the northern Sinai coast of Egypt. Geo Mar Lett 17:140–146

Frihy OE, Deabes EA (2011) Beach and Nearshore Morphodynamics of the Central-bulge of the Nile Delta Coast Egypt. IJEP 1(2):33–46

Frihy OE, Deabes A, El Sayed W (2003) Processes reshaping the Nile delta promontories of Egypt: pre- and postprotection. Geomorphology 53:263–279

Frihy OE, Dean R (1992) Artificial beach nourishment projects on the Egyptian coast. International Coastal Congress (ICC), Kiel, Germany, 7–12 September 1992, pp 84–95

Frihy OE, Fans MA, Khafagy AA, Komar PD (1991) Nearshore sediment transport patterns along the Nile Delta. Egypt. Coastal Eng. 15:409–429

Frihy OE, Khafagy AA (1991) Climatic and human induced changes in relation to shoreline migration trends in the Nile Delta promontories. CATENA 18:197–211

Frihy OE, Lawrence D (2004) Evolution of the modern Nile delta promontories: development of accretional features during shoreline retreat. Environ Geol 46:914–931

Frihy OE (2003) The Nile delta-Alexandria coast: vulnerability to sea-level rise, consequences and adaptation. Mitig Adapt Strateg Strateg Glob Change 8:115–138

Gaweesh KM (2003) Technical report on the study of water level variation along Nile Delta (Progress report No. 1). Coastal Research Institute, Ministry of Irrigation and Public Works

Iskander MM (2013) Wave climate and coastal structures in the Nile Delta coast of Egypt. Emirates J Eng Res 18(1):43–57

Inman DL, Jenkins SA (1984) The Nile littoral cell and man's impact on the coastal zone of the southeastern Mediterranean-Scripps Institution of Oceanography, Reference Series 84–31. University of California, La Jolla, p 43

Kankara RS, Selvan SC, Markose VJ, Rajan BS, Arockiaraj S (2015) Estimation of long and short-term shoreline changes along Andhra Pradesh coast using remote sensing and GIS techniques. 8th International Conference on Asian and Pacific Coasts (APAC 2015). Procedia Engineering, vol 116, pp 855–862

Louati M, Saïdi H, Zargouni F (2014) Shoreline change assessment using remote sensing and GIS techniques: a case study of the Medjerda delta coast, Tunisia. Arab J Geosci. https://doi.org/10.1007/s12517-014-1472-1

Mahapatra M, Ratheesh R, Rajawat AS (2014) Shoreline change analysis along the coast of south Gujarat, India, using digital shoreline analysis system. J Indian Soc Remote Sens 42(4):869–876

Mujabar PS, Chandrasekar N (2013) Shoreline change analysis along the coast between Kanyakumari and Tuticorin of India using remote sensing and GIS. Arab J Geosci 6:647–664

Nandi S, Ghosh M, Kundu A, Dutta D, Baksi M (2016) Shoreline shifting and its prediction using remote sensing and GIS Techniques: a case study of Sagar Island, West Bengal (India). J Coast Conserv 20:61–80

Thieler ER, Himmelstoss EA, Zichichi JL, Ergul A (2009) The digital shoreline analysis system (DSAS) version 4.0—an ArcGIS extension for calculating shoreline change. U.S. Geological Survey Open-File Report 2008-1278 v 4.2

UNDP (2009) Adaptation to climate change in the Nile Delta through integrated coastal zone management. United Nations Development Programme, Project Document 3748, New York, 124p

UNESCO/UNDP (1978) Coastal protection studies. Final Technical Report, Paris

Wang Y, Hou X, Jia M, Shi P, Yu L (2014) Remote detection of shoreline changes in eastern bank of Laizhou Bay, North China. J Indian Soc Remote Sens. https://doi.org/10.1007/s12524-014-0361-0

Chapter 4
GIS-Based Spatial Modeling of Potential Impacts of Sea Level Rise Along the Nile Delta Coast

Abstract The problem of sea level change and coastal flooding is one of the most important environmental problems facing the low-laying coastal zone, not only along the NDC but also along most of the delta's coasts worldwide. Egypt is one of the top five countries in the world threatened by sea level rise by one meter above the current level, especially the NDC (Dasgupta et al., Change 93:379–388, 2009). The Egyptian government has spent 170 million US dollars to protect the Rosetta and Damietta promontories and the Burullus headland from coastal erosion, sea level rise, and storm surges (Iskander, Emirates Journal for Engineering Research 18(1):43–57, 2013). This chapter is divided into three parts: (1) assessing global sea level rise and regional to average sea level, methods of SLR measurement, causes, and future hazards. (2) Related to evidence and geoindicators of sea level change in the Holocene, such as geomorphological evidence of ancient shorelines and sedimentary and geoarchaeological evidence. (3) Application of geographic information systems (GISs) in the spatial modeling of the potential impacts of SLR and land subsidence up to the end of the twenty-first century. The ArcGIS Model Builder interface was used to link spatial multicriteria factors together to compute and predict the future impacts of SLR on human activities and population under different global and regional scenarios.

4.1 General

Sea level changes are typically caused by several natural phenomena, including ocean thermal expansion and glacial melt from Greenland and Antarctica. Global average sea level is expected to rise through the twenty-first century, according to the IPCC projections between 0.18 and 0.59 cm. The Nile Delta is vulnerable due to the impact of climate change and related sea level rise. Because of the low-elevation coastal zone along the NDC, Egypt is considered one of the top five countries expected to be mostly impacted by a 1 m sea level rise resulting from climate change and global warming. Egypt's GDP would be significantly impacted, and Egypt's natural resources, such as coastal zones, water resources, water quality, agricultural land, livestock, and fisheries, may be subjected to vulnerability.

K. S. Darwish, *Hazard Modeling and Assessment of the Nile Delta Coast*,
https://doi.org/10.1007/978-3-031-44324-4_4

In this chapter, GIS-based Model builder and advanced Geo-Processing Spatial analyses in ArcGIS v.10.2 were applied to simulate and assess the potential impacts of coastal flooding due to the Sea Level Rise (SLR) and Land Subsidence along the NDC. The GIS Personal geodatabase was built, including (1) *a topographic dataset* built from digital elevation models (DEMs), which were generated from one arc-second SRTM v.1 data with approximately 30 m spatial resolution and 1 arc-second ASTER- GDEM-V2 data with approximately 30 m spatial resolution and ±1 m (vertical accuracy). (2) *Land-use/cover* classes (water, built-up areas, cropland, wetland, and barren land) were obtained from classified Landsat 8 imagery using maximum likelihood supervised classification and covering the Nile Delta coastal zone. (3) Population density map over the coastal administrative governorates along the NDC (Alexandria, Behira, Kafr El Sheikh, Damietta, and Port Said).

Global sea level rise (SLR) is becoming one of the most concerning environmental issues faced by human society according to the Intergovernmental Panel on Climate Change (IPCC) (Meehl et al. 2007). There are 160 million people currently living in coastal regions that are less than 1 m above sea level (Allison et al. 2009), and even a relatively small magnitude of sea level rise can pose significant threats to human populations and properties close to the coast.

SLR is an ongoing phenomenon that is expected to continue and is projected to have a wide range of effects on coastal environments and infrastructure during the twenty-first century and beyond (Benjamin et al. 2014). Significant changes in climate and their major impacts, such as sea level rise (SLR), are already visible globally. These changes are no longer a distant possibility but a current reality and have become one of the defining challenges for policymakers, industry, and civil society. The consequences of SLR on the coastal areas in Egypt, particularly the Nile Delta, are of major concern to Egypt's population and the government. Warnings from international and regional studies have created a state of national alertness toward the serious implications of SLR (Frihy and El-Sayed 2013). Many of the world's major cities are built in low-lying coastal regions. The impacts of SLR include inundation of these coastal areas, coastal erosion, saltwater intrusion into aquifers, loss of coastal wetlands and mangrove areas, and impacts on biodiversity (Church et al. 2008). Between 1961 and 2003, the global SLR was at an average rate of 1.8 mm/year, and between 1993 and 2003, this rate was higher, averaging approximately 3.1 mm/year (IPCC 2007). According to the United Nations Intergovernmental Panel on Climate Change "IPCC's" scenarios, sea levels will rise by between 18 and 59 cm by the end of the twenty-first century (IPCC 2007). Based on the semiempirical method, the results produce a broader range of SLR projections, especially at the higher end, than outlined in the IPCC Fourth Assessment Report "IPCC AR4." Both the IPCC AR4 and the semiempirical SLR projections are likely to underestimate future SLR if recent trends in the polar region accelerate (Horton et al. 2008; Grinsted et al. 2009) and establish models linking temperature to SLR, projecting twenty-first-century sea level using IPCC projections of temperature, finding that IPCC projections of SLR 2090–2099 are underestimated.

Generally, the Nile Delta is considered to be one of the most vulnerable areas in the world to the risk of SLR. This risk is compounded by varied rates of subsidence movement in different parts of the Nile Delta. This requires careful assessment of the risks associated with expected sea level rise, particularly as the Nile Delta is one of the most populated areas in Egypt and contributes largely to the Egyptian economy. NDC from Alexandria to 35 km east of Port Said has been designated as vulnerable to a rising sea level as a consequence of expected climate changes. The beaches of the delta are partially backed by coastal dunes and brackish water lagoons. Part of the backshore is composed of narrow barriers and low-lying coastal flats (<2 above mean sea level) (Frihy 1992a).

Several studies have revealed that more than 40% of the Nile Delta will be prone to the risk of inundation as a result of SLR by 2050 (Dasgupta et al. 2009; El-Raey 1997; Hereher 2010). These studies have concluded that 0.5 m of SLR could result in the loss of approximately 1,800 km^2 of agricultural land and 12,000 km^2 of urban areas, with more than 3 million people who will be in need of relocation. It is estimated that at 1 m of SLR, the expected loss will be approximately 4,500 km^2 of agricultural land, 25,000 km^2 of urban area, and 24,000 km^2 of wetland, and the inundation risk will threaten a significant portion of the population. Recently, some studies have focused on the potential impacts of SLR on the NDC.

Milliman et al. (1989) assessed the impacts of SLR scenarios ranging between 0 and 217 cm accompanied by land subsidence ranging between 40 and 115 cm by 2100. This mean relative SLR ranges between 28 and 332 cm (Milliman et al. 1989). Based on these SLR scenarios, the study estimated that approximately 21.6–26% of the Nile Delta habitat would be lost due to inundation, which would lead to the displacement of approximately 19–24% of the Delta population.

Frihy (1992a) predicted sea level rise expectations using EPA for scenarios of 0.5, 1.0, 1.5, and 2.0 m sea level rise. The predicted lower and higher sea level rise rates predicted by the EPA (5 mm and 30 mm/yr.) would result in a 2 m rise in sea level by the year 2400 or 2058, respectively. With these rising values, the coastal areas of the western part of Abu Quir Bay, Lake Manzala, and the western part of Tineh Bay might attain maximum land losses of 1.7, 1.9, and 1.4 km, respectively (Frihy 1992a).

In addition, Frihy (1992b) used the trend analysis of sea level rise from the record tide gauges at Alexandria (1944–1989) and Port Said (1926–1987) and indicated a submergence of the land and a rise in sea level of 2 and 2.4 mm/yr, respectively.

Estimates of local future sea level rise by 2100 at Alexandria and Port Said are expected to be 37.9 and 44.2 cm, respectively (Frihy 1992b).

El-Raey et al. (1995a) studied the impacts of SLR on the city of Alexandria under two hypothetical scenarios of 0.5 and 2.0 m. The result was that a one-meter rise in sea level could inundate 1,248 km^2, or 68% of the Governorate. This area contains 95% of existing agriculture, 70% of existing industry, and 50% of existing residential areas and has a present population of approximately 2 million. Sea level rise would also adversely affect the tourist industry. A rise of only 0.5 m could remove nearly all existing beaches. Furthermore, El-Raey (1997) assessed the vulnerability of the coastal zone of the Nile Delta to sea level rise. The results indicate that in these cities alone, over 2 million people will have to abandon their homes, 214,000 jobs and over

$35.0 billion in land value, property, and tourism income may also be lost due to an SLR of 50 cm.

El-Raey et al. (1999) estimated the impacts of SLR over Alexandria and Port Said in particular. The results at Alexandria Governorate indicated that an area of approximately 30% of the city will be lost due to inundation. Almost 2 million people will have to abandon their homeland; 195,000 jobs will be lost, and an economic loss of over $3.5 billion is expected over the next century.

At Port Said Governorate, the results indicated that beach areas are most severely affected, followed by urban areas. The agriculture sector is the least affected sector. It is estimated that the economic loss is over $2.0 billion for 0.50 m SLR and may exceed $4.4 billion for 1.25 m SLR. Furthermore, El Raey et al. (1997) evaluated the vulnerability of the Rosetta area to the impacts of SLR. The study suggested that 72% of Rosetta would be vulnerable to inundation in the case of 1.1 m of SLR, which means loss of all beaches in the area, half of the palm tree cultivation, and threatening 47% of human settlement area.

Moreover, Frihy (2003) assessed the exposure of the Nile Delta coastal zone to SLR consequences. It identified four main sections most vulnerable to SLR, including Burullus and Manzala Lakes, the western backshore zone of Abu Qir Bay, the Manzala-Port Said area, and Ras El Bar beach. The study estimated that approximately 30% of the Nile Delta coast would be vulnerable to SLR. El Nahry and Doluschitz (2010), assuming 1.0, 1.5, and 2.0 m hypothetical rise in sea level, estimated that 28.93, 35.33, and 50.78% of the coastal area of the Nile Delta would be lost under these three scenarios, respectively. Frihy et al. (2010) analyzed the historical records obtained from tide gauges at Alexandria, Rosetta, Burullus, Damietta, and Port Said. It shows a continuous rise in mean sea level fluctuating between 1.8 and 4.9 mm/year; the smaller rate occurs at the Alexandria harbor, while the higher rate occurs at the Rosetta promontory.

Hereher (2010) used remote sensing to assess the vulnerability of the Nile Delta to sea level rise. The analysis of digital elevation models revealed that 18.1% of the delta lies below the mean sea level, 12.7% has an elevation between 0 and 1 m, and 13.1% lies between 1 and 2 m. If the sea level rises by 1 or 2 m, these areas are prone to flooding. The western section of the delta is most vulnerable to a 1 m sea level rise, whereas the entire coastal area is impacted by flooding when the sea level rises by 2 m.

Hassaan (2013) applied GIS-based risk assessment for the Nile Delta coastal zone under different sea level rise scenarios in a case study: Kafr EL Sheikh Governorate, Egypt. The results show that more than 22.59% and 24.50% of the total area of the Kafr El Sheikh Governorate would be vulnerable to inundation under B1 and A1FI (IPCC most optimistic and pessimistic scenarios), respectively. In addition, Hassaan and Abdrabo (2013) indicated that approximately 22.49, 42.18, and 49.22% of the total area of the coastal governorates of the Nile Delta would be susceptible to inundation under different scenarios of SLR.

Moreover, 15.56% of the total area of the Nile Delta was vulnerable to inundation due to land subsidence only, even in the absence of any rise in sea level. Frihy and El-Sayed (2013) showed that beaches and coastal lagoons are highly vulnerable to the potential impacts of SLR and wave overtopping during storm events. The raised coasts will be the least affected by the accelerated SLR. Risk areas in the Alexandria region are Mandara and El Tarh (east of the city), while risk areas in the Nile Delta region are the Manzala Lagoon barrier, east and west of Rosetta City, Gamil, and the Tineh plain.

Zaid et al. (2014) indicated that Mediterranean coastal governorates are ranked as low-moderate and high vulnerability to SLR; over 4 million people will have abandoned their homes from these governorates alone, and 0.5 million jobs and tourism income may also be lost due to SLR. The loss of world-famous historic, cultural, and archeological sites is unaccountable. Finally, Hasan et al. (2015) indicated that 1 m SLR will inundate approximately 580 km^2 (6%) of the total land-cover area, and approximately 887 thousand people will be relocated. Accordingly, the uncertainty analysis of the DEM elevations revealed that the ASTER-GDEM-V2 dataset product was considered the best to determine the future impact of SLR on the Nile Delta region.

Furthermore (El-Raey et al. 1995b), used GIS to assess the vulnerability of various land use and land-cover classes to the impact of sea level rise along the Rosetta area. The results indicated that at 1 m sea level rise, 72% of the total study area could be inundated. This area contains all beaches, half of the palm cultivation, 43% of total urban clusters, which includes 81% of the monumental sites and historic buildings. El-Raey (1997) used GIS and remote sensing techniques together with ground-based surveys to assess the vulnerability of the most important economic and historic centers along the coast, the cities of Alexandria, Rosetta, and Port Said. The results indicate that in these cities alone, over 2 million people will have to abandon their homes, 214,000 jobs and over $35.0 billion in land value, property, and tourism income may also be lost due to an SLR of 50 cm. The loss of world-famous historic, cultural, and archeological sites is unaccountable. The vulnerability of other lowlands in Egypt outside these cities remains to be assessed.

El Raey et al. (1999): Remote sensing techniques and geographic information systems are used for handling spatially referenced data and information to assess the vulnerability of SLR along the Port Said governorate. This study estimated that the economic loss is over $2.0 billion for 0.5 m SLR and may exceed $4.4 billion for 1.25 m SLR. Approximately, 28,000–70,000 people are expected to be displaced, and at least 6,700–16,700 jobs are expected to be lost for the scenario adopted. Hassaan (2013) used GIS to assess risk along the Kafr EL Sheikh Governorate under different sea level rise scenarios. The RSLR and Global IPCC scenarios were used to determine the potential hazards along the Kafr EL Sheikh Governorate until 2100. The results indicate that more than 22.59 and 24.50% of the total area of the Kafr El Sheikh Governorate would be vulnerable to inundation under B1 and A1FI (IPCC most optimistic and pessimistic scenarios), respectively. No significant difference was noticed between the two scenarios in terms of the spatial extent of SLR impacts.

4.2 Coastal Flooding and Inundation Modeling

Inundation models can be used to predict areas that will be flooded based on quantitative relationships between climate and the exposure unit. They can rely on topographic maps to identify potential impact zones, which are identified as those areas that lie below a given elevation contour. These models may be applied at local, regional, or global scales and can address a range of objectives, including predicting sea level rise impacts and raising awareness of these impacts. Potentially inundated areas can be calculated in GIS based on both elevation and proximity to the shoreline. Recently, several global inundation models were designated to visualize and calculate coastal flooding, storm surges, and SLR data at different scales.

4.2.1 SLAAM (Sea Level Affecting Marshes Model)

SLAMM (http://coast.noaa.gov/digitalcoast/tools/slamm) is a GIS-based model developed in the mid-1980 to determine the potential impacts of global climate change on the coasts of the United States. The model was refined and subsequently used in more in-depth studies in Florida, Georgia, Washington, California, and South Carolina. SLAMM was developed specifically for the prediction of different environmental processes affecting wetlands under different scenarios of sea level rise, and SLAMM has the ability to allow marsh migration, resulting in spatial maps that predict cumulative effects on various types of marshes. SLAMM uses a variety of data, including global sea level rise data, NOAA tidal data, detailed wetland data, regional light imaging detection and ranging (LiDAR) data, USGS, DEMs and U.S. National Wetland Inventory data. The model calculates the water height at a specific location using a combination of linear relationships and a tree of decision rules.

4.2.2 Ecological Landscape Spatial Simulation Models

The landscape model (http://ecobas.org/www-server/rem/mdb/btelss.html) was developed to study and predict environmental factors that influence habitat change in wetland habitats in the Barataria and Terrebonne basins on the coast of Louisiana over a period of 30 years. The model connected a hydrodynamic module for flooding over land, using cells of 100 km^2 in size and operating at a time step of one hour, and a spatially articulated ecosystem module, resolving habitat types and changes for 1-km^2 cells in the time step of one hour. The integration of different time and spatial scales was carried out with interpolation routines and mean algorithms. The enforcing functions included dominant regional processes such as subsidence, sedimentation, and sea level rise.

4.2.3 *DIVA Model (Dynamic Interactive Vulnerability Assessment)*

The DIVA Model (http://www.diva-model.net/) is an integrated, state-of-the-art research model of coastal systems that assesses biophysical and socioeconomic consequences of sea level rise and socioeconomic development, taking into account coastal erosion (both direct and indirect), coastal flooding (including rivers), wetland change and salinity intrusion into deltas and estuaries, as well as adaptation in terms of raising dikes and nourishing shores and beaches. DIVA first computes relative sea level rise by combining the sea level rise scenarios with the vertical land movement resulting from glacial-isostatic adjustment and subsidence in deltas.

4.2.4 *SimCLIM Model*

SimCLIM (http://www.climsystems.com) is a software modeling system that simulates, temporally and spatially, biophysical impacts, and socioeconomic effects of climatic variability and change. The system was originally developed by the International Global Change Institute (IGCI, University of Waikato, New Zealand) and is now maintained and distributed by CLIM Systems Ltd., Hamilton, New Zealand. SimCLIM software allows users to generate scenarios of future climate and sea level changes and to examine sectoral impacts or conduct sensitivity analyses. Generally, based on their included datasets, neither SLAMM nor BTELSS are suited for global-scale analyses or for supporting international negotiations and national governments regarding mitigation, adaptation, and policy development. In addition, the DIVA Model is no longer available due to a lack of resources for maintaining and supporting the software, and the SimCLIM model is a commercial tool for which licenses and training courses are available. SimClim ranges in price from inexpensive for a single student license to significantly more expensive for a commercial corporate license and training. If a quick assessment is needed to identify coastal vulnerability to sea level rise (from local to global scales) that requires limited human and financial resources, an inundation model using GIS is the most efficient option. Elevation datasets can be downloaded freely and global sea level rise scenarios.

4.2.5 *GIS-Based Inundation Modeling*

Geographic Information System (GIS) is used to manage and analyze geographical data, process images, produce graphics/maps, model space, and visualize. Due to its ability to analyze spatial data, the GIS approach has been widely used for vulnerability and impact analyses. Many researchers have used GIS for coastal vulnerability assessments. GIS is increasingly used in coastal hazard research for coastal management and information analysis after major storm events due to its ability to evaluate several datasets (Dobosiewicz 2001). GIS-based models are simplified representations of real worlds and their processes. It summarizes the data and makes general statements on the way the phenomena exist and function. GIS models are spatial process models that use spatial data (Lo and Yeung 2007). GIS supports modeling processes in a variety of ways. First, GIS is a tool that uses various data sources, including maps, digital elevation models (DEMs), GPS data, images, and spreadsheets, to process, display, and integrate tables. These data are required for the implementation, calibration, calibration, and validation of the model. Second, models created with GIS can be vector-based or raster-based. The choice depends on the model type, data source, and calculation algorithms (Chang 2014). There are two main methods of spatial modeling sea level rise and subsequent coastal flooding using GIS. Geospatial data represent the real world in two forms, with two different approaches: object-based models and field-based models (Goodchild 1992). The object-based method uses contour lines and is usually suitable for very quick, fast, and easy risk assessment of large areas. The grid model defined by Lo and Yeung (2007) is one of the field-based model variants.

4.3 Global Mean Sea Level Rise

During the geological history of the Earth, the level of the Mediterranean Sea has undergone a number of fluctuations. Researchers' attempts were limited during the first half of the twentieth century, the collection of high coastline remains, dating, and laboratory analysis. Due to the difficulty of collecting data from sunken coastlines, Boring Deep well techniques, echo sounding methods, and radar satellite imaging have enabled researchers to study the characteristics of continental shelves, sunken geomorphological features, river estuaries, and river front deltas. These methods contribute to submerged coastline mapping, dating their age, and mapping ancient positions.

Global mean sea level rise (GMSL) has become one of the most important environmental issues facing human society worldwide in the present and the near future (Meehl et al. 2007).

Approximately 160 million people around the world live in coastal regions less than one meter above sea level (Allison et al. 2009). Therefore, a small increase in sea level may lead to significant threats to populations and their properties in the vicinity of coasts. SLR is defined as an ongoing phenomenon that is expected to persist and predicts widespread impacts on coastal environments and infrastructure during the twenty-first century and beyond (Gutierrez et al. 2014).

4.4 Global Sea Level Measurement Techniques

4.4.1 SLR Measurement Using a Tide Gauge

Tide-gauge records are the most accurate for sea surface data during the last century (Klein and Lichter 2009). Although Cabanes et al. (2001) questioned tidal gauge data and explained them as biased data because they are located in places with abnormal local altitudes in ocean temperatures, which may cause an expansion around the measuring station and give higher than normal measurements. According to Miller and Douglas (2004), tide gauge data are the most precise for sea level research, but measuring sites were rare and unavailable before to 1870. During the twentieth century, the global spread of tidal gauge sensors increased. These data have some disadvantages such as 1) it does not cover all of the world's coastlines; 2) some stations still have just metric data and must undergo RLR processing; and 3) some stations have significant time gaps that have not been recorded (Klein and Lichter 2009).

The Permanent Service for Mean Sea Level (PSMSL) has collected data from the global network of sea level stations recording tide gauges since 1933. The data are received in the form of monthly or annual average values from more than 1,800 stations that record tide data around the world, and the stations send the records in metric format, then PSMSL by rerecording the data to a general comparison surface datum to reduce the monthly and yearly averages with the aim of reducing data correlation with a general comparison surface located 7,000 mm below sea level (the purpose of which is to prevent reading negative values); thus, the values become adjusted to the reference Revised Local Reference (RLR). These data are freely available for all stations around the world (http://www.psmsl.org). To analyze the data in time-series, from which trends in the change in Future Trends.

4.4.2 SLR Measurement Using High-Precision Satellite Altimetry

High-resolution satellites have been monitoring since 1992, and they give measurements covering almost the most of the globe (±66°) with an interval of 10 days (IPCC 2013). Since 1993, three altimeter satellites have been used to measure sea level (TOPEX) (Topography Experiment)/Poseidon, Jason-1, and Jason-2 altimeter measurements need comprehensive corrections in the effects of signals from satellites, atmospheric influence, tidal currents, and other marine effects (Dean and Houston 2013). Since the launch of the TOPEX/Poseidon Altimeter in 1992 (The Gravity Recovery and Climate Experiment [GRACE] in 2002 and the Argo project in 2000), global maps of sea levels have been available to estimate patterns of sea level change with unprecedented accuracy (Church and White 2011).

A report by (IPCC 2007) indicates that during the period 1961–2003, the world average sea level increased by 1.8 mm/year, and during the period 1993–2003, it increased at a higher rate of 3.1 mm/year. The report presented several scenarios for the possible rise of sea level ranging from 18 to 59 cm until the end of the twenty-first century. Horton et al. (2008) and Grinsted et al. (2009) have linked models of temperature to sea level rise based on IPCC (2007) projections for future temperature projections in predicting sea level rise during the twenty-first century. Church and White (2011) estimated that the global average sea level rise between 1880 and 2009 was approximately 210 mm, at a rate of 1.7 ± 0.2 mm/year; from 1961, the rate was 1.9 ± 0.4 mm/year; and since 1900, the rate of increase in SLR has been 0.009 ± 0.004 mm/year.

4.5 Global Sea Level Change

Scientists around the world agree that the global sea level (GSLR) over the past hundred years has risen faster than in the previous two millennia (Douglas 2000).

The global average sea level rise over the past 100 years has been estimated at 1–2.5 mm/year (Brochier and Ramieri 2001), and in the second half of the twentieth century, sea level has risen at a rate of 1.8 ± 0.3 mm/year (Church et al. 2004). Jevrejeva et al. (2014) presented and analyzed tide-gauge tidal data for 1,277 stations worldwide for the period 1807–2009, and the analysis showed the following:

- The sea level increased at a rate of 1.9 ± 0.3 mm/year during the twentieth century and at a rate of 1.8 ± 0.5 mm/year during the period of 1970–2009.
- There is a variation in sea level rise at the regional level, where there is a rapid increase in the rate of sea level rise in Antarctica (4.1 ± 0.8 mm/yr.) since 1970, while the Arctic reached 3.6 ± 0.3 mm/yr.
- The average annual increase in global sea level rise was 0.02 ± 0.01 mm/yr. during the period between 1807 and 2009 (Jevrejeva et al. 2014).
- The rate of sea level change according to satellite measurements gives an average sea level rise of 3.2 mm/yr. for the period 1993–2012, which exceeds the global average of 1.8 mm/yr. for the twentieth century (Jevrejeva et al. 2014).
- According to Cabanes et al. (2001) and Cazenave et al. (2003), Topex/Poseidon satellite data measured sea level rise during the period 1993–2000, estimated at a rate ranging between 2.5 ± 0.2 mm/year.
- A comparison of the rate of sea level rise from tide-gauge records of 456 stations around the world during the time period from 1993 to 2011 with satellite data for the same time period shows that the average global sea level rise was 3.26 mm/year for tide-gauge data and 3.09 mm/year for satellite data (Dean and Houston 2013).

4.6 Factors Controlling Sea Level Change

Mimura (2013) summarized the most important factors controlling sea level change based on studies of Cazenave and Llovel (2010) and Church et al. (2001) as follows:
Factors Related to the Volume of Water in Ocean Basins Include:

- Thermal expansion of seawater.
- Growth/shrinkage of glaciers and ice caps.
- Growth/shrinkage of ice sheets such as the Greenland ice sheet (gis), the west Antarctic ice sheet, and the west Antarctic ice sheet (WAIS).
- Groundwater storage, dam reservoirs, lakes, and groundwater depletion.

Factors Associated with Changes in the Shape and Size of Oceanic Basins:

- Glacier Isostatic Adjustment (GIA).
- Tectonic Movements Including Ground Subsidence.
- Ground Uplift Movements Associated With Earthquakes Landing And Ground Lift Associated With Ground Pressure And Groundwater Pumping.
- Sediment Inflow from the Ground.

Other Factors

- Changes in Ocean Currents.
- Atmospheric Pressure Changes.
- Tidal Currents, Tsunamis, Storm Surges, and Waves.
- Annual Natural Variations such as the Pacific Decadal Oscillation (PDO).

4.6.1 Impact of Climate Change and Global Warming on Sea Level Change

The effects of climate change have emerged globally on sea level rise (SLR), and these changes are no longer distant but have become a confirmed reality and have become one of the most important challenges facing decision-makers, industry, and civil society. In addition, the threat of sea level rise to the sinking of many major global cities in low-lying coastal regions will also cause increased rates of coastal erosion, saltwater intrusion in soil and groundwater reservoirs, and the loss of large areas of wetlands plus the threat to coastal biodiversity (Church et al. 2008). IPCC (2007) indicates that eleven years (1995–2006) were among the warmest years in the record of global surface temperature since (1850).

The linear trend for the hundred-year period (1906–2005) of the rate of increase in global temperatures in the fourth report AR4 was 0.74 (0.56–0.92) °C, which is greater than the corresponding trend of 0.6 (0.4–0.8) in the third report (IPCC, 2001, AR3) for the period (1901–2000). The increase in temperature is widespread throughout the world, and the rate of increase is greater than that in the upper northern latitudes.

Since 1961, the global average sea level has increased at an average rate of 1.8 (1.3–2.3) mm/year and an average rate of 3.1 (2.4–3.8) mm/year since 1993, contributing to the increase in the rate of sea level rise due to thermal expansion, melting glaciers, ice caps, and polar ice sheets. Data collected by satellites since 1978 show that the average annual ice area in the Arctic Sea has shrunk by 2.7 (2.1–3.3)% per decade, with a further shrinkage in the summer of 7.4 (5–9.8)% per decade, while mountain glaciers and snow cover have seen an average decline in both hemispheres (IPCC 2007). Current sea level rise is one of the major indicators of global climate change (IPCC 2013). The two main global contributors to sea level rise are the thermal expansion of ocean waters due to warming oceans, the increase in ocean mass due to the melting of land ice, and the storage of terrestrial water and dams (Cazenave and Remy 2011). Global warming caused by increased concentrations of greenhouse gases in the atmosphere, such as carbon dioxide, is the main contributor to sea level rise (Church et al. 2007).

4.6.2 Impact of Global Ice Melting on Sea Level Change

The thermal expansion of the oceans contributed to the rise in sea level in the period between 1955 and 1995 at a rate of 0.4 mm/year and less than 25% of the measured height during this period, while in the period 1993–2003, the contribution of thermal expansion of ocean waters to sea level rise was estimated at a greater rate of 1.6 mm/year and by 50% of the measured sea level rise of 3.1 mm/year. Glaciers and ice caps (GICs) are the most important contributors to the current rise in global mean sea level, and refrigerators and ice hoods (GICs) (except Greenland and Antarctica) lost an estimated amount of ice (–148 ± 30 gigatons/year) from January 2003 to December 2010, estimated to contribute to sea level rise (0.41 ± 0.08 mm/year) (Jacob et al. 2012).

The contribution of Greenland and Antarctic ice sheets, including some margins of glaciers and ice caps (GICs) (–384 ± 71 Gtn/year), to sea level rise (1.06 ± 0.19 mm/year) in the period (2003–2010) is estimated (Jacob et al. 2012). The total contribution of all ice-covered regions losing an average of ice (–536 ± 93 Gt/year) to sea level rise (1.48 ± 0.26 mm/year) confirms a direct relationship between melting ice and sea level rise (Willis et al. 2010).

4.7 Potential Impacts of SLR

The impacts of sea level rise on coasts are well known, as sea level rise leads to coastal inundation as a result of increased flooding and saltwater intrusion in coastal aquifers and contributes to increased coastal erosion rates. Coastal wetlands such as saltmarshes and mangroves are vulnerable to landslides unless they receive sufficient sediment supplies to protect them from sea level rise (Nicholls and Cazenave 2010).

Sea level rise affects coastlines unevenly, with impacts varying from region to region and time period to time (Nicholls et al. 2007). The SLR impact is great on low-level coasts such as deltas, coastal plains, islands, and near coastal urban areas (Williams 2013).

Roads and urban centers located in low-lying areas close to the coast are also at risk of frequent flooding during periods of spring high tide and storms (Douglas 2001). The ongoing SLR consequences of sea level rise in Egypt, especially on the NDC, threaten both residents and the government. Warnings from preliminary findings of global and regional research have created a state of awareness among populations toward the serious implications expected from sea level rise (Frihy and El-Sayed 2013). Sea level rise affects beach decline, as it is one of the most fundamental factors in historical changes in beaches (Mohanty et al. 2017). Continuous coastline retreat is caused not only by coastal erosion but also by submergence of low-laying areas (Bird, 1996). The IPCC (1998) indicates that a one centimeter vertical rise in sea level could retreat the shoreline by one meter horizontally. Bruun (1962, 1988) introduced the famous equations linking the relationship between shoreline retreat, increasing sea level rise, and shorefront slope, as evidenced by the following equation (Mohanty et al. 2017):

$$Sh.r = \mathrm{SlrT}/\tan(\beta)$$

where, sh.r = rate of shoreline retreat, SlrT = rate of sea level rise, β = shore slope angle.

Frihy (1992a) studied the response of shoreline retreat and possible sea level rise, and several scenarios were applied (0.5 m, 1 m, 2 m) using the Bruun equation (1962) on 55 coastal strip beach profiles distributed on the shores of the study area. It was concluded that the shoreline will retreat a distance of 1.9 km, 1.7 km, and 1.4 km in the areas of Lake Manzalah, western Abu Qir Bay, and the Tienah coastal plain, respectively, if sea level rises 2 m by 2,200 m.

4.8 Geoindicators of SLR During the Holocene

There is much evidence and many indicators of sea level change in the study area, the most important of which are as follows.

4.8.1 Geomorphological Indicators

Mapping ancient shorelines is considered one of the most important indicators for coastline dynamics and sea level change during historical times. It used archaeological remains and sites to identify ancient sea level changes. Paleo shorelines are important geomorphological evidence in the study area because there are many indicators of ancient shorelines, and these indicators include ancient sand dunes, terraces, and the slope of the continental shelf. It can be used to assess the relation between sea level change and coastline dynamics along the NDC (Frihy 1992a). Sea level fluctuation during the Holocene affected the sediments and their morphology along the continental shelf. The observations confirm the maximum sea level altitude and provide basic information about the morphology of the continental shelf, which generally consists of a series of terraces (50–90 m, 40–50 m, 15–27 m depth) separated by easy slopes cut by submerged channels, and these terraces are related to the Flandrian transgression that occurred at intervals starting from 14,000 to 15,000 years and 12,500 to 10,000 years (Misdorp and Sestini 1976). The fluctuation in sea level change during the Holocene affected the topography of the NDC, as shown in Fig. 4.1.

Geomorphological indicators show that sedimentary sandy beach edges are signs of coastline retreat during the late Holocene. They are beach sediments that have grown parallel to the current shoreline on the eastern side of the Rosetta Promontory, Burullus, Damietta, and in the coastal plain of Tina. These sand ridges appear as remnants of ancient beach sediments from the Ancient Transgression, and this is confirmed by the fact that Tell El-Farma, which was located at the mouth of the Old Paleozoic branch, 30 km east of Port Said, is currently located at a distance of more than 2.3 km inside the ground away from the coast. These sand ridges consist of fine to coarse sand grains and shell fractures (Stanley and Toscano 2009).

Evidence of the maximum advance of the ancient shoreline has been drawn from the positions of the ancient dunes located south of the coastal lagoons, and these dunes

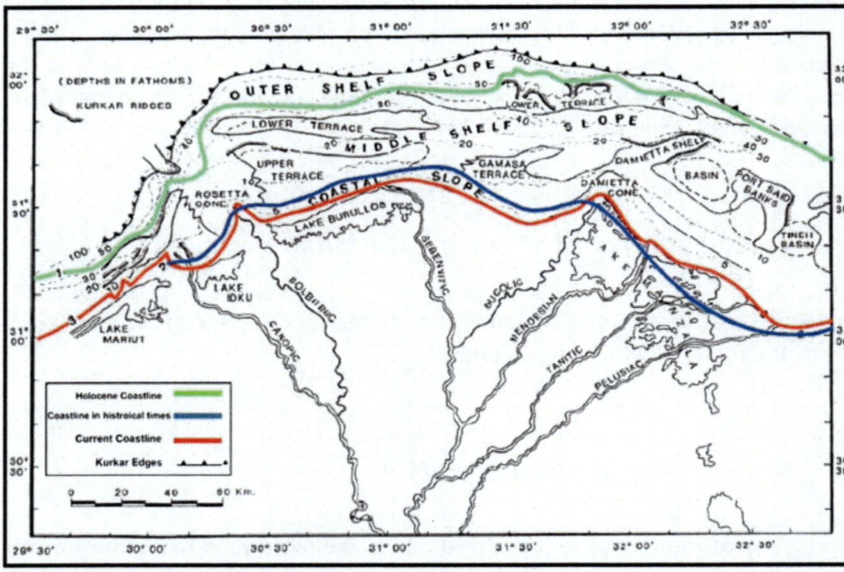

Fig. 4.1 Ancient shoreline along the NDC during the Holocene, after Said (1981)

date back to the late Holocene and are no more than 3,000 years old. The geographical distribution of the ancient dunes south of the Delta lagoons confirms that the source of the sand is derived from the nearby shores of these dunes. Logically, the ancient beaches must have existed before the formation of the Delta lakes. According to many studies, the lakes were formed approximately 961 AD by maritime advances, probably in the context of sea level rise. This evidence suggests that the maximum progress extended near the positions of ancient sand dunes south of the margins of present-day coastal lakes. This line of shore is located on the ancient branches of the Nile, and the sites of ancient branches of the Nile have been identified from historical information, marine morphology, and sedimentary forms (Coutellier and Stanley 1987). Said (1981) indicates that the Nile Delta coastline was at a distance extending in a northerly direction from the current position by approximately 40–50 km and approaching the depth line of 100 m, and this line is found at the same average depth for all continental shelves estimated by (Shepard 1973). It expresses the lowest sea level during the period of the Wurm Regression, as shown in Fig. 4.1.

4.8.2 Sedimentological Indicators

According to Summerhayes et al. (1978), the texture of the Nile Delta continental sediments consists mainly of terrigenous sands located on the slope of the Nile Delta, terrigenous muds located on the Rosetta and Damietta sides, and calcareous muds located on the upper continental shelf. Sestini (1989) suggests that the remaining coarse sand along the shore north of Burullus lagoon and west of Abu Qir Bay is expected to be mixed with pre-Holocene sands rather than with sand brought from the estuaries of the Nile during historical periods. This distribution of sand deposits confirms that the delta coastline retreated during the Holocene. Submersion due to sea level rise during the Holocene has been documented by deep core sediment profiles in the northeastern Nile Delta (El Askary and Frihy 1986; Coutellier and Stanley 1987). The lithofacies sedimentary sequence consists of the Prodelta Mud and the Delta-front Delta and formed over the period 9000–3000 years, and the sedimentary sequence in the ancient and anterior deltas occurred mainly during the time when the rate of sea level rise diminished to the greatest degree (Frihy 1992c), as shown in Fig. 4.2.

4.8.3 Archaeological Indicators

Archeology is a useful technique for determining the direction and rate of relative change in sea level, especially in the Mediterranean, coasts that have been sites of human habitation and reconstruction for thousands of years (Emery et al. 1988). Most of the low-lying coasts of the world's fluvial-marine deltas were nonglaciated and relatively tectonically stable during the Holocene but were subject to both sea

Fig. 4.2 Vertical cross-section of the NDC as an indicator of SLR (along Al Gharbeya Drain Inlet)

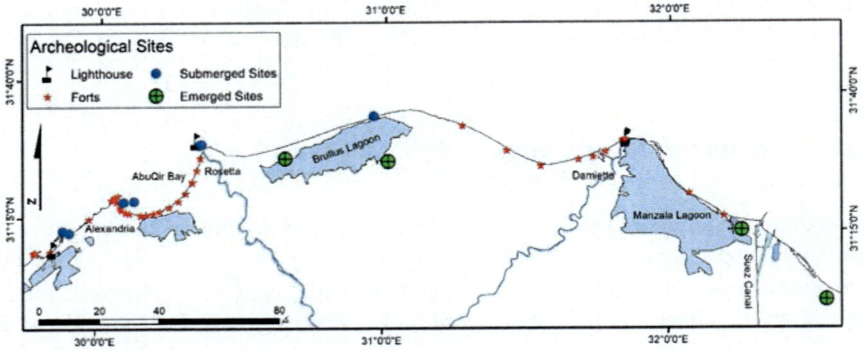

Fig. 4.3 Archaeological sites along the NDC

level rise and subsidence. As a result, the urban settlements that are located on the coastal margins of the large deltas, or close to them in historical periods, may at the present time be partially buried in sediment or completely submerged under the sea (Stanley and Toscano 2009). Figure 4.3 shows the archaeological sites along the NDC.

4.8.3.1 The Sunken Harbors

Roman and Islamic remains are distributed along the backshore of Alexandria, Burullus, the islands of Manzala lagoon, and the Tinah coastal plain, most of which have been submerged or emerged under sediments as a result of sea level rise and fall during historical periods (Frihy 1992c). There are seven coastal sites (submerged, and emerged) dating from the Middle to Late Holocene (5400–4700 years BP) to approximately 1,400 years (archaeological age) and distributed between northwest to northeast of the margins of the Nile Delta; three of them are located under the sea (submerged) called the eastern port of Alexandria (three sunken ports at depths of –

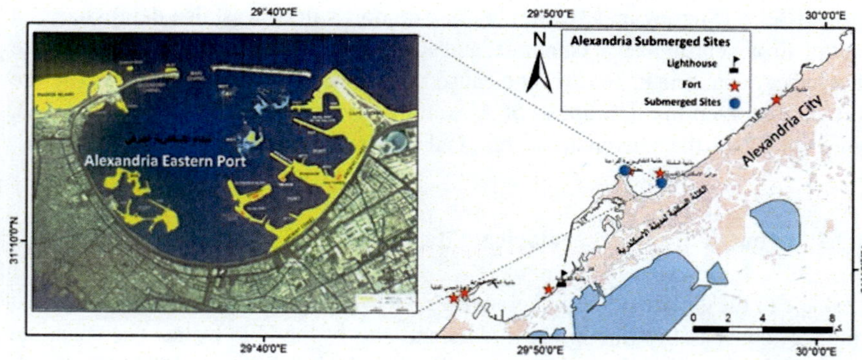

Fig. 4.4 Submerged archeological sites in Alexandria eastern port. After Goddio (1998)

6.5, −7.5, −13 m), east of the canopy (−6 m), Heracleon (−8 m); and the other four sites are buried in the delta sediments: the site northwest of Lake Burullus (+1 m), southeast of Lake Burullus (+1 m above sea level), Tell Tenis (+1 m), Tell Al-Belouzi "Mouth of the Peluzioc Branch" (+5.5 m) (Stanley and Toscano 2009).

Alexandria Eastern Port

Several archaeological explorations of the post-Alexander period (Ptolemaic, Roman, Byzantine, and Islamic) from 332 BC to 642 CE were revealed by underwater surveying on the bottom of the eastern port of Alexandria (Goddio 1998). The dating of newly collected sediments from the eastern port provides direct evidence of human activities from the periods preceding Alexander the Great in 332 BC (Stanley et al. 2007). Sediment dating sectors have revealed numerous anthropogenic materials ranging in age from 3000 to 2332 years and consisting of ceramic tiles, rock fractures brought from central and southern Egypt, high percentages of heavy metals and organic rocks in deep sediments (−12 m below current sea level). These diverse components represent an important record of human activities for the seven centuries preceding the arrival of Alexander the Great's at the coastal margins (Stanley and Toscano 2009), as shown in Fig. 4.4.

Above these ancient pre-Alexander remains of walls and jetties, temple columns, and the remains of a newer seaport. These remains were built after the division of the Gulf of Alexandria into two separate ports by the Greeks and Romans 2,100 years ago (Hesse 1998). Remains of sunken islands, peninsulas, and piers of three seaports, all sunken in the east and southeast port of Alexandria, were also uncovered (Goddio 1998).

The divers discovered near the port shore the recent remains of ceramics dating back to the Byzantine and Islamic eras, oil bulbs, and other remains at a depth of −5.5 m from the current sea level, and most of the large sunken forms in the port of Alexandria were sunk during the late Roman and Islamic periods due to sea level rise and landslides (Stanley and Warne 1998). The port's floor was also depressed as a result of strong earthquakes, tsunamis, and strong sea storms in winter, such as those that struck Alexandria in 365 AD and caused significant damage to the infrastructure and that destroyed the lighthouse of Alexandria in 1323 AD (Stanley et al. 2006). Figure 4.4 shows the sunken archaeological sites in the eastern port of Alexandria.

4.8.3.2 AbuQir Bay Archaeological Site

The shape of the Gulf of AbuQir has a clear concave shape toward the sea, in contrast to the soft convex coastline in the middle and east of the Nile Delta. The coastline is approximately 50 km from the western border of AbuQir Bay (AbuQir headland) to the Rosetta Estuary, and the Abu Qir headland represents the northern part of the bay extending 5 km in the direction of the sea from the modern edge of the AbuQir promontory. Wilson Island is one of the remaining islands of the ancient AbuQir headland. The Canopic branch is one of the largest branches of the ancient Nile Delta and flows northwest of the Nile Delta toward the Mediterranean Sea. The historical documents, geographical and geological evidence, and archaeological explorations in the Gulf confirm that the Canopic branch was active in the historical period from 4000 BC until the end of the first millennium AD, where the river fluvial discharge of the branch continued during the Predynastic periods, dynastic periods, Greek periods, Roman periods, and Byzantine periods (Stanley et al. 2004). The Canopic channels flow at two different sites: Heraklion and Eastern Canopus, and these sites are submerged because of oceanic changes in sea level, annual flooding, fault growth, and other marine processes (Stanley et al. 2004), as shown in Fig. 4.5.

These sunken urban settlements have been explored in the Abu Qir Bay area 5 km north of the coast and at a depth of 5–7 m. These cities were originally built by the Greeks in the Canopic bay for trade in Egypt, and their locations give a good view of the evolution of the coast and the ancient branches (Goddio 2004). The ancient Ruins remains of the city of East Canopy were identified by underwater archaeological survey, geophysical sectors, and analysis of deep samples collected using Vibracoring technique, and it is located at a distance of 1.8 km southeast of the current city of Abu Qir (ancient Abu Qir known as Canopus) and is located at a depth of (−6 m) and an area of 3 km^2 and among the most important remains that have been discovered: jewelry, coins, and wall debris. This city was constructed in the pre-Ptolemaic era and was originally built on a Cemented Limestone from the Pleistocene period, and this Headland defines the western margins of the Gulf (Stanley and Toscano 2009).

The city of Heraklion has sunken in the Gulf of Abu Qir; it was located at a distance of 3.9 km on the eastern side of the city of East Canopy, covering an area of 1.5 km^2 of the sea floor, buried with sediments 2 m thick, and located at a depth of 6 to 7 m below current sea level. This site is rich in a variety of archaeological remains, including walls, temples, and artifacts (Goddio 2007). According to this geoarchaeological

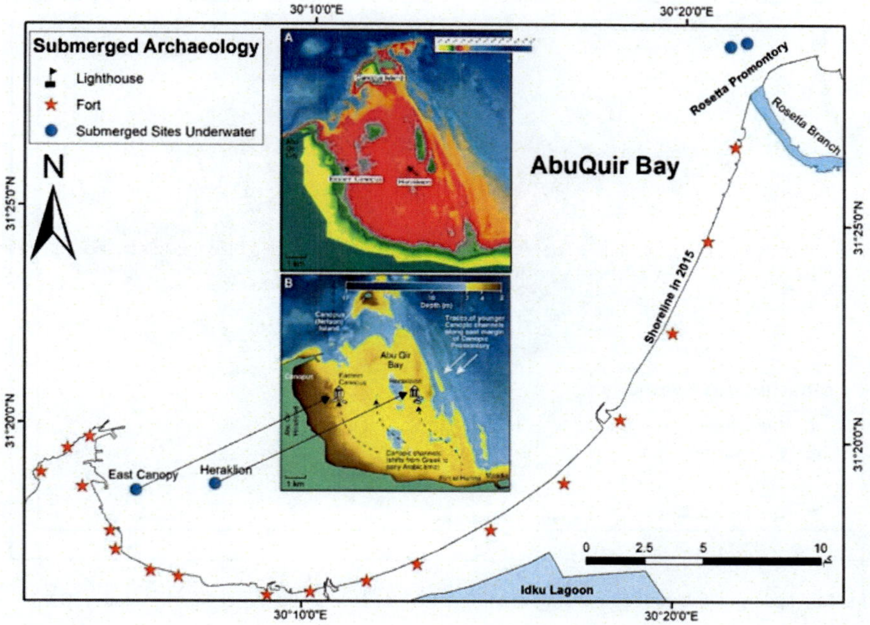

Fig. 4.5 Submerged archeological sites in AbuQir Bay. After Stanley et al. (2004)

evidence, over the last 1,000 years, the coastline has retreated to its present position in the Abu Qir Bay area for up to 5 km southerly and approximately 2 km westward from the ancient sites uncovered (Stanley et al. 2004).

4.8.3.3 Archaeologically Emerged Sites

The site of the ancient port of Pelusium is one of the most important archaeological sites buried under the surface of the Delta deposits. Buried sites due to the advance of the coastline seaward, where the "Fortified Port of Pelusium" is a city and a commercial center, were later known as Tel Farama, which is currently located at a distance of approximately 3 km south of the current coastline at a level of +0.5 m. Most of the explorations during the last century indicated the large archaeological remains at this site back to the Late Roman and Byzantine eras (Stanley and Toscano 2009). The archaeological remains are found at a depth of up to 5.5 m from the current sea level, including some circular grained bricks dating back to 2700 BC, and this city was built during the 22nd and 23rd dynasties (Stanley et al. 2008). Stanley (1990) believes that the mouth of the Paleozoic branch extended south of the current coastline by approximately 10 km and formed a thick sedimentary cone of flood formations, as shown in Fig. 4.6.

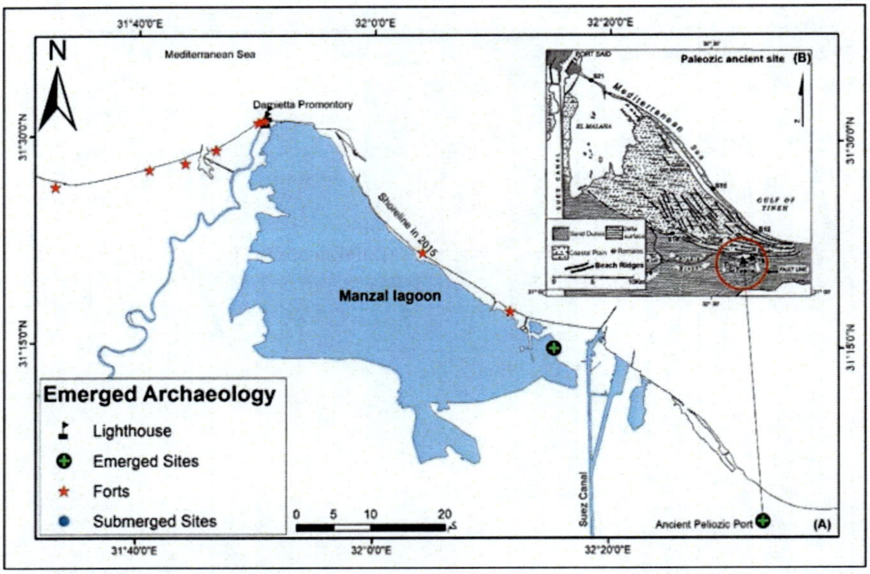

Fig. 4.6 Emerged archaeological sites along the eastern Nile Delta. After Stanley et al. (2008)

4.9 GIS-Based Empirical Modeling of Potential Impacts of SLR

The GIS-based spatial model is defined as a simplified representation of the real world and its operations. It summarizes the data and presents the general components of the existence and procedures of the phenomenon. Thus, it is considered a model for a spatial process in which spatial data are employed (Lo and Yeung 2007). One of the most essential elements in GIS model building is identifying the nature of the model, the modeling process, and the role of GIS in the modeling process (Chang 2014).

The creation of a GIS model requires a set of organized steps: (1) define the objectives of the model, in which the problem of the study and the phenomenon will be created and the spatial and temporal scales are identified. (2) Divide the model into a group of elements to determine the characteristics of each element and the interaction between the elements in the form of a conceptual diagram using a mathematical model with a set of mathematical equations and commands. (3) Model calibration aims to calibrate and implement the model and check the validity of the data.

GIS technology can help to process, display, and integrate data from different sources, such as maps, digital elevation models (DEMs), GPS data, satellite visualizations, and tables. These data are necessary to calibrate and examine the validity of the model's construction and running (Chang 2014). Spatial models in a GIS environment may be vector-based models or raster-based models, where the choice depends on the nature of the model, the source of the data, and the calculation computing algorithms.

- The variation between vector and raster models does not prevent Model Builder from linking data for both in the modeling process (Chang 2014).
- The modeling process may occur entirely within the GIS environment or require linking with some other computer software, and the most important GIS programs that contain extensive analytical functions for modeling are ArcGIS, GRASS, IDRISI, ILWIS, MF-works, and PCRASTER.
- There are two ways to build a spatial model of sea level rise and coastal flooding using geographic information systems, where spatial data visualize the real world in two forms, the object-based model and field-based model (Goodchild 1992). The object-based model is used for rapid assessment of simple risk over a large area, while the field-based model is best representative of the spatial phenomenon that persists over a large area, such as the cellular model (Lo and Yeung 2007).
- The GIS model is graphically represented by a diagram to facilitate the creation, visualization, editing, and execution of geoprocessing workflows for use, reuse, deployment, and application to different geographic areas (Schaller and Mattos 2009). GIS spatial models make modeling easier, completing workflows by linking tasks and processes together, implementing, modifying, and iterating workflows multiple times with a single mouse click, facilitating workflow management, and increasing the efficiency of geoprocessing (Goodchild 2005).
- The ModelBuilder graphic user interface is an application used to create, edit, and manage forms. Models are workflows that are linked together by a series of geoprocessing tools where the output feeds the first process tool Output to enter the second process as Input. The ModelBuilder is also known as a visual programming language for building workflows (Allen 2011), and within a form builder, a form consists of one or more processes.
- The process consists of input data, spatial functions to apply operations to inputs, and output data. The processes are linked together so that the outputs of the first operation are inputs to the second operation, and the model does not contain spatial data but places that contain spatial data called nodes that display the data to which the operations are applied and that arise during the application of the Model Running model (ESRI 2000).

According to Allen (2011), model elements are the building blocks of models, and the most important elements of the model are as follows:

- ***Tools***: Basic spatial operations for building form templates and workflows within a form and performing geographic and tabular operations.
- ***Variables***: These elements of the model contain a value or reference to the data stored on the computer's memory. There are two types of variables: data variables and value variables.

- ***Connectors***: Arrows that automatically link data and values to tools. The links indicate the direction of the operations, and there are four types of links: data links, model environment links, precondition links, and feedback links, as shown in Fig. 4.8.

This study uses ESRI ArcGIS Model Builder technology to build a geospatial model for assessing the potential impacts of SLR and land subsidence along the Nile Delta coastal zone in different scenarios. A diagram that facilitates the creation, visualization, editing, and execution of geoprocessing workflows, their use and reuse, and their sharing and application to different geographic areas graphically represents the GIS model.

ModelBuilder is an application that is used to create, edit, and manage models. Models are workflows that string together sequences of geoprocessing tools, feeding the output of one tool into another tool as input. ModelBuilder can also be thought of as a visual programming language for building workflows. The ModelBuilder innovation allows complex processes to be displayed in the form of flowcharts in a graphical user interface. The user can plan the work to be carried out, which tasks to perform and in which sequence, and then automate the workflow by easily stringing processes and tools together in the model diagram. Thanks to the diagram, every step of the process can be traced and documented on the fly. Datasets (inputs and outputs) and geoprocessing tools can be selected by drag-and-drop into the model window and connected graphically with arrows according to the processing sequence. A model process consists of a tool and all variables connected to it. Connector lines indicate the sequence of processing (Figs. 4.7, 4.8, 4.9).

Model elements are the basic building blocks of models. There are three types, as shown in Fig. 4.9:

1. **Tools**: Geoprocessing tools are the basic building blocks of workflows in a model. Tools perform various operations on geographic or tabular data. When tools are added to a model, they become model elements.
2. **Variables**: Variables are elements in a model that hold a value or a reference to data stored on disk. There are two types of variables:

   - Data: Data variables are model elements that contain descriptive information about data stored on disk. Properties of data that are described in a data variable include field information, spatial reference, and path.

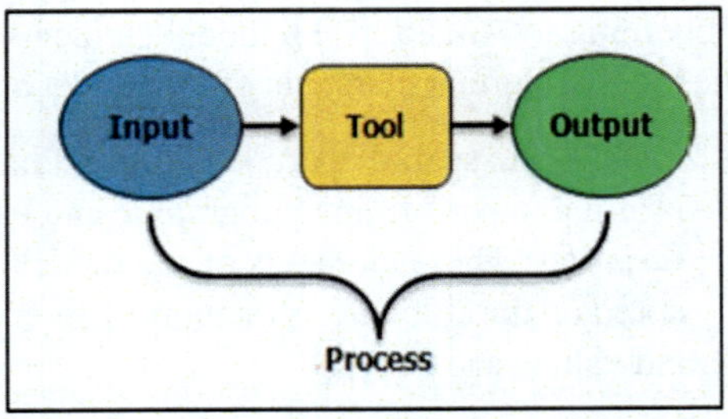

Fig. 4.7 Simple model process. After https://desktop.arcgis.com/en/desktop/latest/analyze/modelbuilder/understanding-process-state.htm

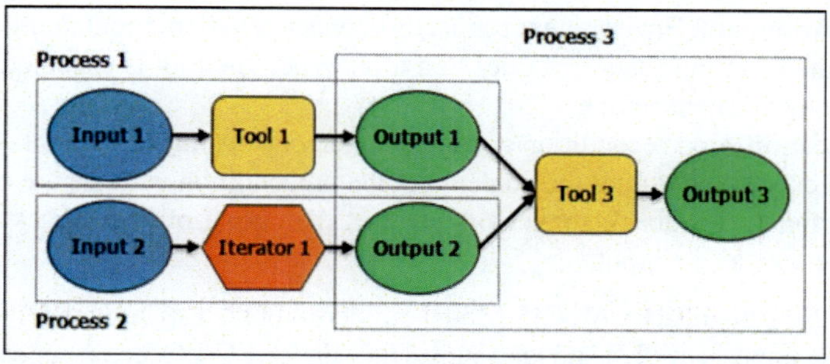

Fig. 4.8 Several processes in a model. After https://desktop.arcgis.com/en/desktop/latest/analyze/modelbuilder/understanding-process-state.htm

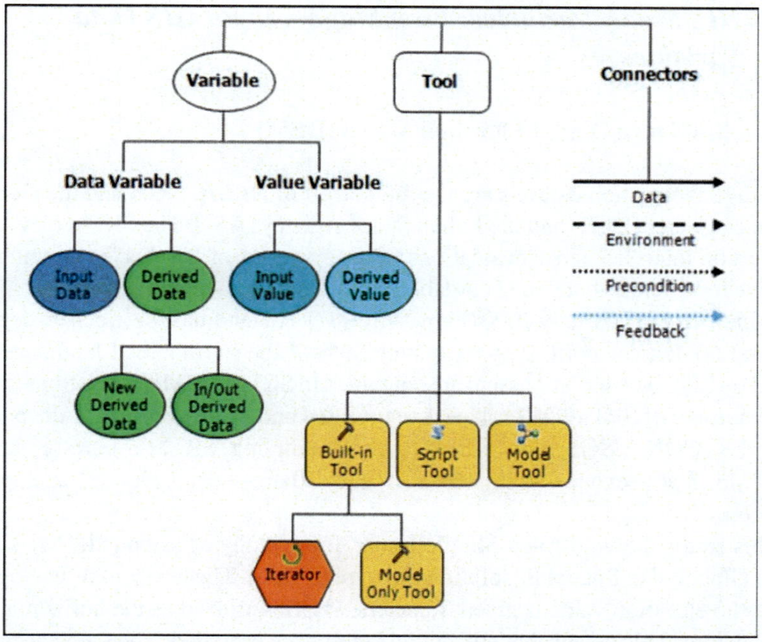

Fig. 4.9 The diagram shows model elements as classified in ModelBuilder. After https://desktop.arcgis.com/en/desktop/latest/analyze/modelbuilder/model-elements.htm

- Values: Value variables are values such as strings, numbers, Booleans (true/false values), spatial references, linear units, or extents. Value variables contain anything but references to data stored on disk.

3. **Connectors**: Connectors connect data and values to tools. The connector arrows show the direction of processing. There are four types of connectors:

- **Data**: Data connectors connect data and value variables to tools.

- **Environment**: Environment connectors connect a variable containing an environment setting (data or value) to a tool. When the tool is executed, it uses the environment setting.
- **Precondition**: Precondition connectors connect a variable to a tool. The tool will execute only after the contents of the precondition variable are created.
- **Feedback**: Feedback connectors connect the output of a tool back into the same tool as the input.

This study uses the ESRI's ArcGIS Model builder to build a spatial model to assess the potential impacts of SLR and land subsidence on the LU/LC features, population, and main roads along the Nile Delta coast by the end of this century.

4.9.1 GIS Model Variables: First Stage—Input GIS Data Variables

4.9.1.1 Building a Digital Elevation Model (DEM)

SRTM digital elevation datasets are the joint effort of NASA, NGA and the German Aerospace Center (DLR) and the Italian Space Agency (ASI). The SRTM elevations are based on interferometric evaluations of observations of the dual radar antennas (sensitive for C-and X-band) on board the Shuttle Radar Topography Mission's spacecraft, which flew in February 2000 (Farr et al. 2007). All landmass between 56 degrees south and 60 degrees north (approximately 80% of the Earth's total landmass) are covered by SRTM observations and are contained in SRTM DEM observations. Since 2000, a number of SRTM DEMs have been created and made available to the public, initially by 168 the USGS, with different ground sampling (SRTM1: 1 arc-second/30 m; SRTM3: 3 arc-seconds/90 m; 169 SRTM30: 30 arc-seconds/900 m) and spatial coverage.

In this study, the void-filled Shuttle Radar Topography Mission (SRTM) 1 Arc-Second Global with Spatial Resolution (30 m pixel size). The horizontal spacing is 1 arc sec; the elevation value is given in meters. WGS84 is used as the horizontal and vertical datum. This means that ellipsoidal heights are provided. The DEM accuracy requirements are ±16 m absolute and ±6 m relative vertical accuracy (Rabus et al. 2003). Prasad (2016) indicated that the absolute vertical accuracy of SRTM DTED-2 (30 m) (1 arc-second) is <18 m, and the relative point-to-point vertical accuracy is <12 m. The accuracy of the DEM is highly important for applications based on the three-dimensional model of the land. The accuracy of the DEM varies according to the position accuracy, frequency, distribution of the source data used, land structure, and DEM generating (interpolation) methods (Ebaid 2014).

Table 4.1 and Fig. 4.10 show the technique applied to build the digital elevation model in this study. ESRI's ArcGIS Model Builder was used to execute geoprocessing in several steps: (1) Seven SRTM scenes covering the Nile Delta region were mosaicked. (2) Reproject Mosaic into a metric projection UTM, WGS, Zone 36.

(3) Fill gaps and remove null values. (4) Generalization of DEM using Focal Statistics. (5) Clip Study Area Boundaries. (6) Extract simplified contour lines with 1-m intervals, as clearly seen in Fig. 4.11.

Table 4.1 Specifications of the SRTM dataset used in this study

SRTM Entity Id	Acquisition date	Resolution	Vertical accuracy	Covering	Published
SRTM1N31E029V3	11-FEB-00	1-ARC	<18 m	Alexandria	23-SEP-14
SRTM1N30E029V3	11-FEB-00	1-ARC		Al-Hamam	
SRTM1N31E030V3	11-FEB-00	1-ARC		Rosetta	
SRTM1N30E030V3	11-FEB-00	1-ARC		Southern	
SRTM1N31E031V3	11-FEB-00	1-ARC		Rosetta	
SRTM1N30E031V3	11-FEB-00	1-ARC		Damietta	
SRTM1N31E032V3	11-FEB-00	1-ARC		Middle Delta	
SRTM1N30E032V3	11-FEB-00	1-ARC		Port said	
				Suez Canal	

Source http://earthexplorer.usgs.gov/

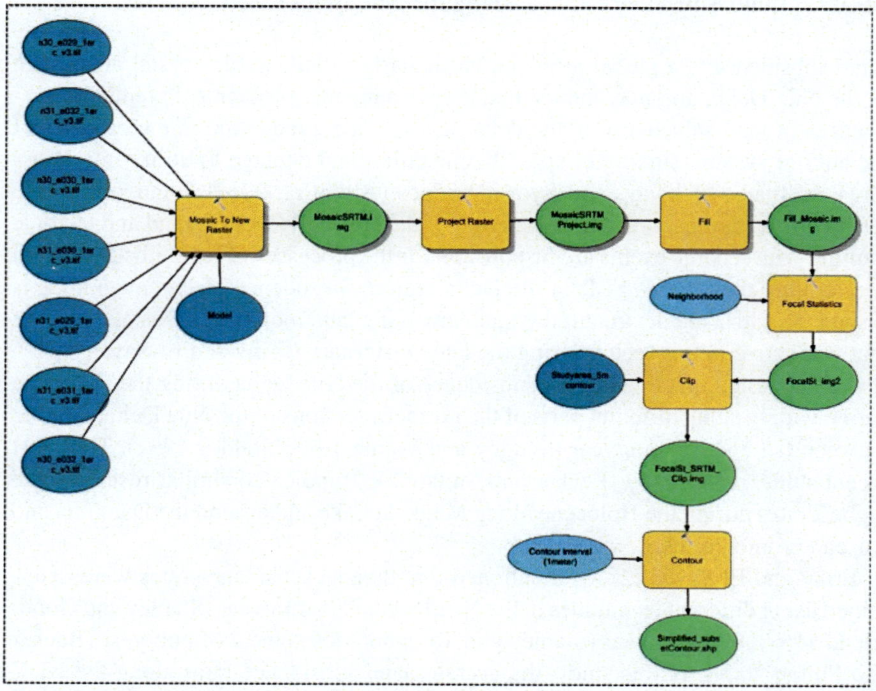

Fig. 4.10 Building DEM from SRTM v.1 using ArcGIS Modelbuilder

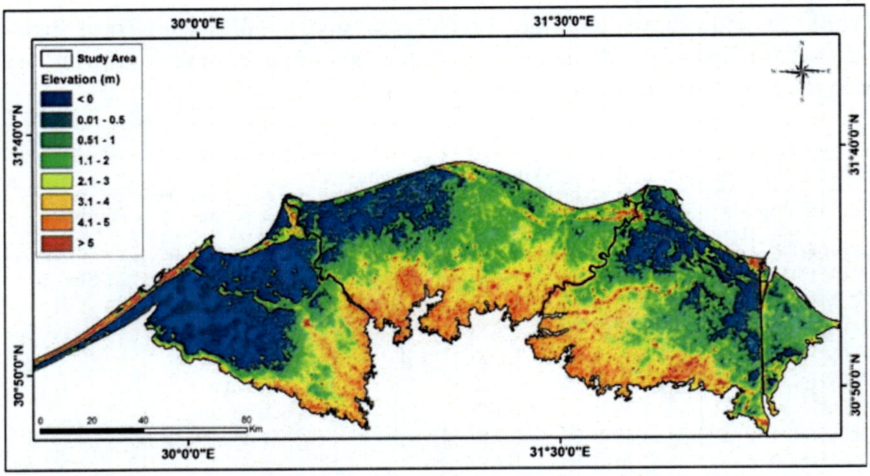

Fig. 4.11 Topography of the Northern Nile Delta Coast Derived from SRTM, v.1 arc-second

4.9.1.2 Land Subsidence Rates Along the Nile Delta Coast

Land subsidence is a global problem, particularly in vulnerable coastal areas such as the Nile Delta, and may induce major environmental problems. Potential consequences of land subsidence in the delta may include a reduction in the capacity of the aquifer system, sinking of coastal zone cities, and damage to utility infrastructures, railroads, highways and flyover bridges (Aly et al. 2012). Land subsidence resulted from compaction and dewatering of the Holocene Deltaic mud and tectonic faulting with seismic events are responsible for the processes of land subsidence and accelerating the relative SLR in the Delta region. Subsidence of this magnitude is associated with a major structural system involving fault motion and isostatic loading that was active in this sector during the late Quaternary (Frihy and El-Sayed 2013). It was suggested that the rates of subsidence of the Nile Delta during the Holocene varied widely along different parts of the northern sections of the Nile Delta, ranging between 0.5 and 4.5 mm/year (Stanley and Warne 1993; Stanley 1997). The most recent study, reported by (Becker and Sultan 2009) indicated similar results of the highest rates during the Holocene along Manzala Lake and around the Damietta and Rosetta promontories.

Frihy and El-Sayed (2013) mentioned that diverse subsidence rates were determined using different techniques in the Nile Delta: 3.98 mm/year (Stanley and Goodfriend 1997), 4–5 mm/year (Stanley and Toscano 2009), and 2–8 mm/year (Becker and Sultan 2009). In this study, the average land subsidence rates along the NDC ranged between 0.5 and 5 mm/yr. for the past 7,500 years) calculated from 87 samples of radiocarbon-dated core sediments from deep wells in the northern delta plain (Stanley and Warne 1993; Stanley 1997; Frihy and El-Sayed 2013). The scan of the hard copy map was converted to a vector layer and then converted to a raster layer compatible with the model variables 30 m pixel size, as shown in Fig. 4.12.

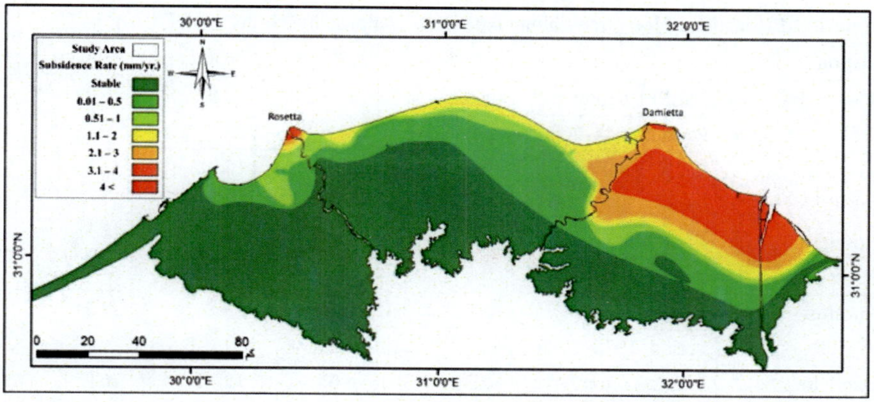

Fig. 4.12 Land subsidence rates along the NDC based on Stanley and Warne (1993) and Stanley (1997)

4.9.1.3 Relative Sea Level Rise (RSLR)

The vulnerability of coastal areas to inundation due to SLR can typically be determined by local topography in relation to expected relative SLR. Global SLR (Eustatic Changes in Sea Level), land subsidence, and gross accretion of fluvial sediment (Ericson et al. 2006) in turn determine such relative SLR. Expected relative SLR at any point of time in the future can be calculated according to Eq. 4.1 (Hassaan 2013).

$$R_{SLR} = (G_{SLR}) + (V * T) - S \tag{4.1}$$

where: $\boldsymbol{R_{SLR}}$ = Relative sea level rise; $\boldsymbol{G_{SLR}}$ = Global sea level rise; **V** = Annual rate of land subsidence; **T** = Number of years between present and future year of concern; **S** = Gross accretion of fluvial sediments.

However, due to the construction of the High Dam in Aswan during the 1960s, all the sediments were trapped in the reservoir. Accordingly, the gross accretion of fluvial sediments (S) in the case of the Nile Delta coastal zone is nearly equal to zero (Frihy et al. 1996). Hence, the expected relative SLR at any point in time in the future in the Nile Delta coastal zone can be calculated according to the previous equation without the gross accretion of fluvial sediments (S), as in Eq. 4.2 (Hassaan 2013).

$$R_{SLR} = (G_{SLR}) + (V * T) \tag{4.2}$$

Table 4.2 Estimation of R_{SLR} rates along Nile Delta stations, after Frihy et al. (2010)

Station		2015/cm	2030/cm	2050/cm	2100/cm
Alexandria	*RSLR*: 1.8 mm/yr	1.62	4.32	7.92	16.92
	Period: 1944–2006				
Abu Qir	*RSLR*: 3.4 mm/yr	3.4	8.5	15.3	32.3
	Period: 1992–2005				
Rosetta	*RSLR*: 4.9 mm/yr	3.43	10.78	20.58	45.08
	Period: 1969–2008				
Burullus	*RSLR*: 3.0 mm/yr	2.7	7.2	13.2	28.2
	Period: 1975–2006				
Damietta	*RSLR*: 2.9 mm/yr	2.32	6.67	12.47	26.97
	Period: 1990–2007				
Port Said	*RSLR*: 2.8 mm/yr	4.2	8.4	14	28
	Period: 1926–2000				
Average NDC	*RSLR*: 3 mm/yr	2.9	7.6	13.9	29.6
	Period:1926–2008				

4.9.1.4 Sea Level Rise Along the Nile Delta Coast

Long-Term SLR

The Permanent Service for Mean Sea Level (PSMSL) published online Revised Local Reference (RLR) data taken from the observed data of the tide-gauge registering along the coasts of the world. Egypt has two stations (Alexandria ID: 503) with data from 1945 to 2006. In addition, Port Said station (ID: 253) has data from 1923 to 1946. As the absolute sea level change is difficult to measure, the study of Frihy et al. (2010) estimated that the relative sea level rise (R_{SLR}) was derived from tide-gauge data relative to fixed tide-gauge benchmarks. The analysis of the long-term records of the tide-gauge data was measured along five stations along the Nile Delta. The annually averaged tide-gauge records were calculated from the hourly heights of sea level taken (with discontinuance) at Alexandria (1944–2006; 60 years), AbuQir (1992–2005; 14 years), Rosetta (1969–2008; 17 years), Burullus (1976–2006; 24 years), Damietta (1990–2007; 13 years), and Port Said harbor (1926–2000; 47 years) (Frihy et al. 2010). The updated sea level changes determined along the Nile Delta coastal plain were linked with the geological processes acting in this study area (Thickness of Holocene sediments, subsidence rate, and tectonic activity). The results indicated that the mean sea level at Alexandria, Abu Qir, Rosetta, Burullus, Damietta, and Port Said has risen at annual rates of 1.8, 3.4, 4.9, 3.0, 2.9, and 2.8 mm/year, respectively. According to the previous results of Frihy et al. (2010), the R_{SLR} at the end of this century 2100 along Alexandria, AbuQir, Rosetta, Burullus, Damietta, and Port Said will be 16.92, 32.3, 45.08, 28.2, 26.97, and 28 cm, respectively, as shown in Table 4.2, Figs. 4.13, and 4.14.

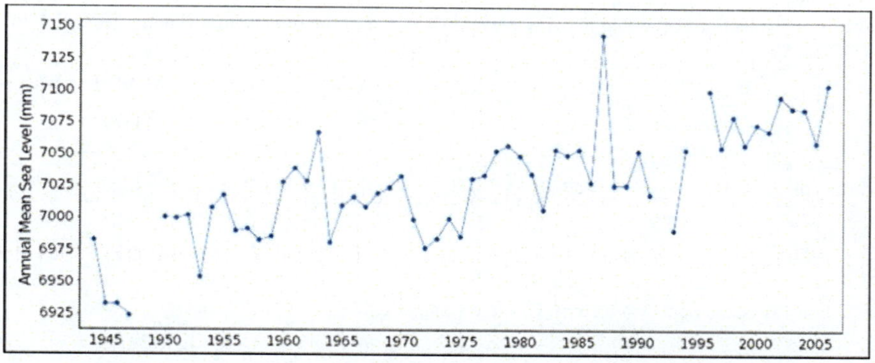

Fig. 4.13 Annual mean sea level rise in Alexandria station since 1945 (*Source* http://www.psmsl.org)

Frihy et al. (2010) attempted to calculate the acceleration of SLR along Alexandria station by analyzing the annual records to assure acceleration of sea level rise during the twentieth century up to the present time (1944–1979 and 1980–2006 periods). The results show minor statistically significant acceleration of R_{SLR} compared to trends estimated between 1944 and 1979 (34 years) and 1980–2006 (26 years). In general, the curve of 1980–2006 appears to be slightly higher than the 1944–1979 trends. This indicates that sea level at this particular site has been accelerating during the last 26 years. This could be attributed to the active episodic effects of land-level changes, emergence or subsidence, and because of the effect of global warming. Evidently, the Alexandria coastal plain has been affected by diverse vertical land motion, emergence, and subsidence, which resulted from the effect of the AbuQir and Rosetta fault zones (Frihy et al. 2010). A sea level increase of 3.0 mm/year results from average rates estimated at Alexandria, AbuQir, Rosetta, Burullus, Damietta, and Port Said. These average rates fit within the lower estimated prediction made by the

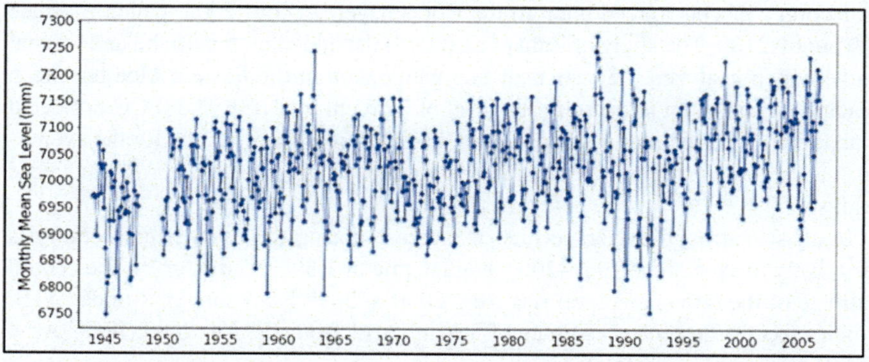

Fig. 4.14 Monthly mean sea level rise in Alexandria station since 1945 (*Source* http://www.psmsl.org)

Table 4.3 Projections of global mean sea level by the end of the twenty-first century according to IPCC (2007)

		B1		B2		A1B		A1T		A2		A1FI	
		Optimistic Scenario										(Worst Scenario)	
Mean Sea Level Rise	m	0.18	0.38	0.20	0.43	0.21	0.48	0.20	0.45	0.23	0.51	0.26	0.59
	mm yr^{-1}	1.5	3.9	2.1	5.6	2.1	6.0	1.7	4.7	3.0	8.5	3.0	9.7

IPCC Fourth Assessment Ar4 Report: Climate Change 2007

Table 4.4 Projections of global mean sea level by the end of the twenty-first century according to IPCC (2013)

			Optimistic			Worst
		SRES A1B	RCP2.6	RCP4.5	RCP6.0	RCP8.5
Mean S L R	m	0.60 [0.42 to 0.80]	0.44 [0.28 to 0.61]	0.53 [0.36 to 0.71]	0.55 [0.38 to 0.73]	0.74 [0.52 to 0.98]
	mm yr^{-1}	8.1 [5.1 to 11.4]	4.4 [2.0 to 6.8]	6.1 [3.5 to 8.8]	7.4 [4.7 to 10.3]	11.2 [7.5 to 15.7]

IPCC Fifth Assessment Ar5 Report: Climate Change 2013

Intergovernmental Panel on Climate Change (IPCC). The IPCC's Third Assessment Report projected that the global average sea level would rise by between 9 and 88 cm between 1990 and 2100 (Frihy et al. 2010). Recent studies, which assessed the potential impacts of SLR along the Nile Delta coast, used different scenarios: Hassaan (2013) depended on B1 (28 cm) and A1FI (43 cm) scenarios, and through 2100, the IPCC, SRES scenarios 2007. In addition, the study of Hassaan and Abdrabo (2013) was based on the IPCC A1FI scenario, Rahmstorf, Pfeffer 49, 140, 200 cm through 2100. The study of Frihy and El-Sayed (2013) depended on the global climate projection concepts agreed upon in the Copenhagen Accord 2009, which estimate 100 cm by 2100. The study of Zaid et al. (2014) depended on the Micha and Michal study, which analyzed the statistical tide-gauge data in the Eastern Mediterranean Sea and estimated an increase in sea level of 22.5 cm by 2100. The IPCC projected changes in global average sea level under the SRES marker scenarios for the twenty-first century due to thermal expansion and land ice changes based on AR4 AOGCM results.

In all scenarios, the average rate of increase during the twenty-first century is very likely to exceed the 1961–2003 average rate of 1.8 ± 0.5 mm yr^{-1}. The central estimate of the rate of sea level rise from 2090 to 2099 is 3.8 mm yr^{-1} under A1B, which exceeds the central estimate of 3.1 mm yr^{-1} from 1993 to 2003 (IPCC Ar4 2007). As shown in Tables 4.3, 4.4.

According to the IPCC, Fifth assessment report 2013: (1) global sea level is rising, (2) this rise has accelerated since preindustrial times and (3) it will accelerate further

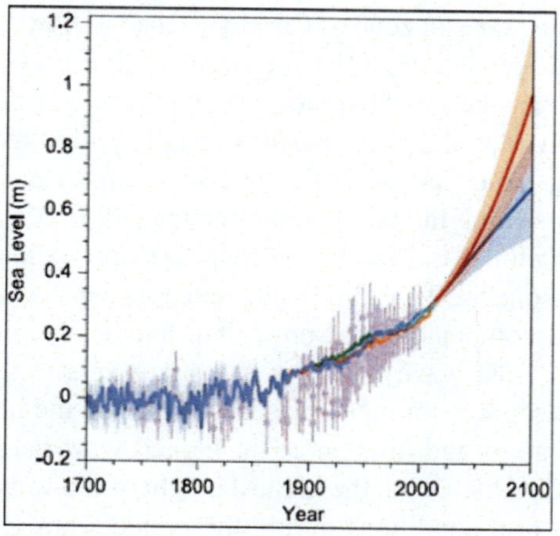

Fig. 4.15 Acceleration of mean SLR according to IPCC, Fifth assessment report 2013

in this century. The projections for the future are much higher and more credible than those in the fourth report but possibly still slightly conservative, as we will discuss in more detail below. For high emissions, the IPCC now predicts a global rise of 52–98 cm by 2100, which would threaten the survival of coastal cities and entire island nations. Nevertheless, even with aggressive emissions reductions, a rise by 28–61 cm is predicted. Even under this highly optimistic scenario, we might see over half a meter of sea level rise, with serious impacts on many coastal areas, including coastal erosion and a greatly increased risk of flooding (IPCC, Ar5, 2013) (Fig. 4.15).

In this study, the IPCC Global Sea Level Rise (GSLR) scenarios in the 2013 Ar5 assessment report, RCP 2.6 An optimistic scenario [0.28 to 0.61] with an average of 0.44 m and RCP 8.5 as the worst scenario [0.52 to 0.98] with an average of 0.74 m will be used to assess the vulnerability of the Nile Delta Coast to the GSLR and its potential impacts on land use/land cover.

Short-Term SLR

A sudden rise in sea level and marine inundation occurs because of extreme events resulting from sea storm waves and tsunami-seismic waves, which may cause coastal disasters and inundation of coastal cities and lowland coastal zones, the most important of which are as follows:

Storm Surges

High-intensity storms cause submersion seawater for part of the city, and a combination of factors affecting the strength of their impact are wave height, beach topography, Topo bathymetry, sea level, and coastal protection (AASTMT 2011). Frihy et al. (2010) indicated that strong sea storms occurred in December 2003 in Alexandria city and resulted in high and strong waves that caused the submersion of the low coastal zone of Corniche Road (2.4 m) in the Mandara area and damaged it. On December 27, 2010, the coast of Alexandria was struck by the strongest storm surge over the last 50 years, reaching a height of seven meters, and it had a significant impact on the sea barriers installed to protect the beaches (Soliman et al. 2014). In general, the effect of wave action and storm surge in this study area is seasonal in its strength and direction (Nafaa 1995). Storm waves begin in the winter from mid-October to March, and there are approximately sixteen storms occurring annually along the NDC, including seven strong storms with a range of 2–4 days called Nawat, accompanied by strong winds and heavy rain (Iskander 2013).

The wave height in the study area is seasonal; it increases in the winter and decreases in other seasons. Waves of the north-coast west of Alexandria are more energy and stronger, as the highest wave height reached 7.72 m, while on the coast of the Nile Delta, the highest height of the waves reached 6.02 m, which occurs during storm conditions during the period from 1982 to 2000 m (Frihy et al. 2010). The height of storm waves from (N–NNW) directions has been recorded with an average of (2–5 m), while storm waves from (NW–NNE) directions are less than 2 m (Frihy et al. 2008).

Deabes (2003) estimated the height of storm surges over three areas along the NDC based on annual, monthly, and daily storm data.

- The highest height of storms in the AbuQir area ranged between −51 and 47 cm. Along the Rosetta promontory, the height of the storms ranged between −46 to 45 cm and −35 to 59 cm along the mouth of the Rosetta promontory.
- Storm height ranged between −55 and 47 cm in the Damietta area and between −40 and 33 cm along the Damietta promontory.
- The topography of the beaches plays an important role in sea level change. The average height of storms along the NDC ranges from −40 to 40 cm, with the highest height being 59 cm and the lowest height being −55 cm.

Tsunami

Many devastating tsunami events caused by earthquakes have been recorded along the NDC, which has affected the city of Alexandria since the first century AD. Alexandria city is more threatened by tsunamic waves because of the passing of major fault lines close to it, including the Suez-Cairo-Alexandria fault line and Qattara-Eratosthenes fault line. According to the global achievement of the registered tsunami data in the Mediterranean by the National Geophysical Data Center (NGDC). The city of Alexandria was exposed to only two tsunami-seismic earthquakes that led to mass destruction on July 21, 365 AD, and August 8, 1303 AD (Papadopoulos and Fokaefs 2005). Records indicate that more than 50,000 houses were destroyed by the tsunami in Alexandria in 365 AD, killing more than 5,000 people in the city, and the island of Crete was completely destroyed. In 1303 AD, earthquakes destroyed the island of Rhodes in Greece, followed by a strong tsunami wave that affected the city of Alexandria and North Africa (Eckert et al. 2012). According to the previous events, the frequency of tsunamis that hit the Nile Delta was estimated at 1/800 years (the last confirmed tsunami was 1303 AD); in the event of their occurrence, at least there are several scenarios to affect the study area and the city of Alexandria, which are +5 m (best-case scenario) and +9 m (worst-case scenario). From the application of the scenarios to the Al-Jumrok and Al-Montazah areas in Alexandria, it is clear that in scenario (5 m), 12% of the buildings in the Al-Jumrok and Al-Montazah areas are located in the very dangerous area, and the impact will be limited to the Al-Montazah area, which would be in the high-risk range (Eckert et al. 2012).

4.9.1.5 Land-Use/Cover (LU/LC) Categories

Remote sensing is considered an important and effective technique for detecting land cover with the ability to provide easy and simple information regarding the spatial variations of the Earth's surface. Land cover refers to the type of material present on the landscape (water, crops, forest, wetland, human-made materials such as asphalt, e.g.). Land use refers to what people do on the land surface, such as agriculture, commerce, and settlement. e.g., (Jensen 2015). Remote sensing has long been an important and effective means for monitoring land cover with its ability to quickly provide broad, precise, impartial, and easily available information regarding the spatial variability of the land surface. The particular remote sensing data source is an important factor for successful land-cover classification. Landsat satellite data are commonly used remote sensing data for land-cover classification (Gumma et al. 2011; Gong et al. 2013). The availability of global datasets from Landsat has the potential to significantly improve the characterization of the Earth's land surface (Kun et al. 2014). Compared to Landsat 7, Landsat 8 has several new features: (1) two sensors, OLI and TIRS, (2) two spectral bands that include coastal aerosol and cirrus bands, (3) one thermal band that is split into two bands, (4) a refined spectral range for some bands, improving the spectral responses across the channels, near-infrared (NIR) and panchromatic bands, (5) improved radiometric resolution from 8 to 12 bits (Pahlevan and Schott 2013), and (6) the change in instrument design has resulted in significant improvements in signal-to-noise ratios (SNR), almost twice as good as Landsat7 (Irons et al. 2012). According to the USGS land-use/land-cover classification categories, Level I LU/LC information can be obtained from Landsat satellite imagery data or from high-altitude imagery and can also be interpreted from conventional large-scale aircraft imagery or compiled by ground surveys. This same information can be displayed at a wide variety of scales ranging from a standard topographic map scale, such as 1:24,000 or even larger (Anderson et al. 1976). Landsat 8 OLI satellite imagery taken in 2015 scenes numbers (P177r38, P176r38, P177r39, P178r39) were preprocessed, enhanced, color matched, mosaicked, subset to study area, and classified into several classes as shown in Fig. 4.16.

Satellite digital image classification techniques mean grouping pixels to represent land-cover features. Land cover could be forested, urban, agricultural, and other types of features. There are three main image classification techniques. Pixels are the smallest unit represented in an image. Image classification uses the reflectance statistics for individual pixels. Unsupervised and supervised image classification

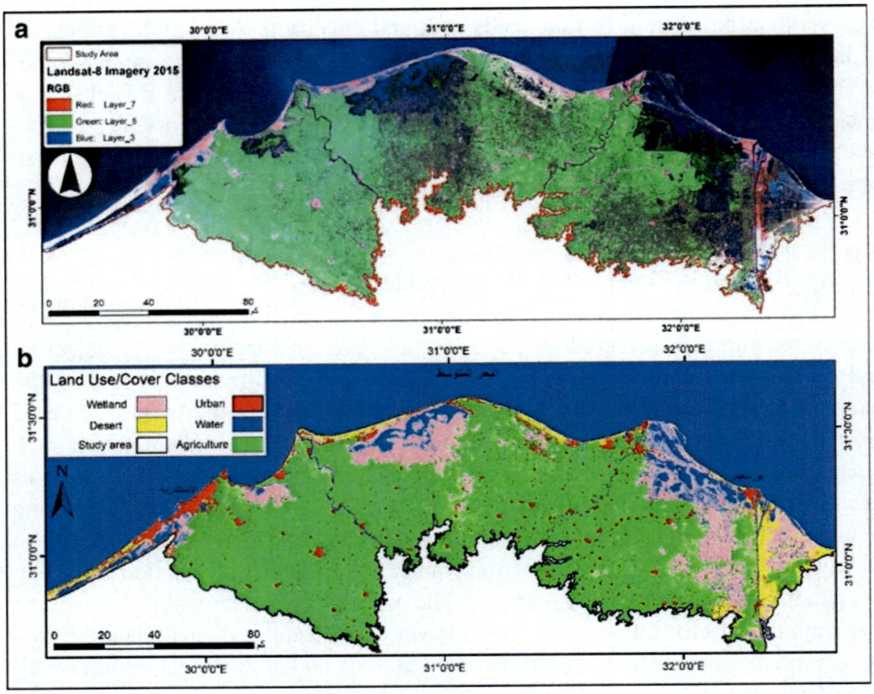

Fig. 4.16 Landsat OLI imagery taken in 2015 along NDC; **a** Landsat Mosaic of NDC, and **b** Supervised classification of land-use/cover categories

techniques are the two most common approaches. However, object-based classification has been breaking more ground as of late. In supervised classification, the user selects representative samples for each land-cover class in the digital image. These sample land-cover classes are called "training sites." The image classification software uses the training sites to identify the land-cover classes in the entire image. The classification of land cover is based on the spectral signature defined in the training set. The digital image classification software determines each class on what it resembles most in the training set. The common supervised classification algorithms are maximum likelihood and minimum-distance classification. In this study, more than 3000 training sites were selected as signatures in the areas of interest (AOI) to classify five different land-cover classes (urban and built-up areas, agricultural land, water, wetland, and barren land):

- **Urban and built-up areas**

Urban or built-up land comprises areas of intensive use with much of the land covered by structures. Included in this category are cities, towns, villages, strip developments along highways, transportation, power, and communications facilities, and areas such as those occupied by malls, shopping centers, industrial and commercial complexes, and institutions that may, in some instances, be isolated from urban areas (Anderson et al. 1976).

- **Agricultural Land**

Agricultural land may be defined broadly, as land used primarily for the production of food and fiber. On high-altitude imagery, the chief indications of agricultural activity will be distinctive geometric field and road patterns on the landscape and the traces produced by livestock or mechanized equipment. However, pasture and other lands where such equipment is used infrequently may not show as well-defined shapes as other areas (Anderson et al. 1976).

- **Water**

The delineation of water areas depends on the scale of data presentation and the scale and resolution. Characteristics of the remote sensor data used for interpretation of land use and land cover. Water as defined by the Census Bureau includes all areas within the land mass of the United States that are persistently water covered, including streams and canals, lakes, reservoirs, bays, and estuaries (Anderson et al. 1976).

- **Wetland**

Wetlands are areas where the water table is near or above the land surface for a significant part of most years. The hydrologic regime is such that aquatic or hydrophytic vegetation is usually established, although alluvial and tidal flats may not be vegetated. Wetlands are frequently associated with topographic lows, even in mountainous regions. Examples of wetlands include marshes, mudflats, and swamps situated on the shallow margins of bays, lakes, ponds, streams, and manmade impoundments such as reservoirs. They include wet meadows or perched bogs in high mountain valleys and seasonally wet or flooded basins, playas, or potholes with no surface-water outflow. Shallow water areas where aquatic vegetation is submerged are classed as open water and are not included in the Wetland category (Anderson et al. 1976).

- **Barren Land**

Barren land has a limited ability to support life, and less than one-third of the area has vegetation or other cover. In general, it is an area of thin soil, sand, or rocks. Vegetation, if present, is more widely spaced and scrubby than that in the shrub and brush category of rangeland. Extreme events, such as heavy rainfall, occasionally result in the growth of short-lived, more luxuriant plant cover, including dry salt flats, beaches, sandy areas other than beaches, bare exposed rock, strip mines, quarries, and gravel pits (Anderson et al. 1976). ERDAS Imagine v. 2014 was used to perform supervised classification. The maximum likelihood (MLC) and support vector machine (SVM) classifiers were selected for the land-cover classification of OLI data. The MLC has been the most popular parametric classifier used for remote sensing data classification (Kun et al. 2014).

- **Post-Classification Accuracy Assessment**

Table 4.5 Post-classification accuracy assessment matrix

		Ground truth					No. classified pixels	%
		Urban	Water	Agriculture	Wetland	Barren land		
Classified in Satellite Image as:	Urban	30	0	2	0	2	34	88.2
	Water	3	77	0	1	0	81	95
	Agriculture	4	0	72	11	1	88	82
	Wetland	0	9	3	56	1	69	81
	Barren land	5	0	0	0	23	28	82
No. Ground Truth Pixels		42	86	77	68	27	300	86

The users of LULC maps need to know how accurate the maps are to use the data more correctly and efficiently (Smits et al. 1999).

In this study, to assess the accuracy of the Landsat 8 classification for study area land use/land cover, 300 random points were distributed along the original Landsat imagery. The true pixels were identified using Google Earth imagery and saved in a shapefile with spatial coordinates. The points were converted from points to raster with the same spatial projection and pixel size as the classified thematic imagery. The predicate pixels were automatically collected using the Combine technique. Dbase was converted to a table, and then the matrix was calculated (Table). The results of the classification accuracy were correct at 86%. According to Anderson et al. (1976), the minimum level of interpretation accuracy in the identification of land use and LU/LC categories from remote sensing data should be at least 85% (Table 4.5). The most widely promoted classification accuracy is in the form of an error matrix, which can be used to derive a series of descriptive and analytical statistics (Manandhar et al. 2009).

$$\text{Theoverall accuracy} = \text{correct prescriptions}/\text{total prescriptions}$$

4.9.1.6 Population Density Map

Arab Republic of Egypt Census statistical data for the year 2006 done by the Egyptian Central Agency for Public Mobilization and Statistics (CAPMAS) were used. The cartographic distribution of population density was made to all coastal governorates and administrative divisions, which are eight governorates: Alexandria, Beheira, Kafr El Sheikh, Dakahlia, Sharkia, Damietta, Port Said, and Ismailia, as shown in Fig. 4.17.

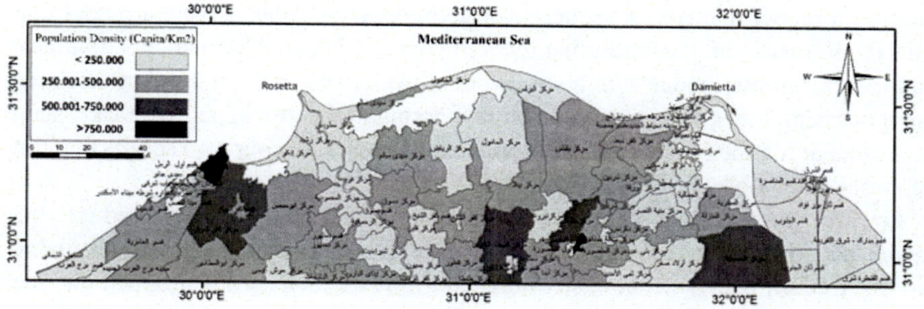

Fig. 4.17 Population density map along the coastal governorates

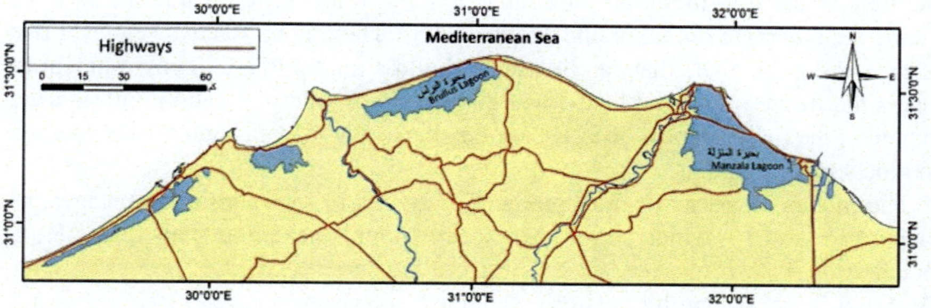

Fig. 4.18 Road network covering the coastal zone of the NDC

4.9.1.7 Highway Network

Highways represent the only means of transportation in the study area. An updated main coastal road network was digitized using 50,000 topographic maps and Google Earth images, as shown in Fig. 4.18.

4.10 Model Functions, Validation, and Running

Geoprocessing is the processing of geographic information, one of the basic functions of a geographic information system (GIS). It provides a way to create new information by applying an operation to existing data. Any alteration or information extraction you perform on your data involves a geoprocessing task. It can be a task such as converting geographic data to a different format, or it can involve multiple tasks performed in sequence, such as those that clip, select, and then intersect datasets. Within ArcGIS®, geoprocessing tasks can be performed in a number of ways. (1) Run a tool using its dialog box. Open a tool's dialog box, fill in parameter values, and click OK to run the tool. (2) Run tools at a command line. Type the tool name and its parameter values at a command line, and then press Enter to run the tool. (3) Build and run a model that runs a sequence of geoprocessing tools. Create a model that runs the sequence of geoprocessing tools in your workflow. Alter parameter values, and then rerun the model with a single click. (4) Create and run a script that runs geoprocessing tools. Use system batch processing scripts for repetitive tasks, such as those that run the same tool on multiple inputs, or create your own scripts that run geoprocessing tools (ESRI 2004).

Operations are defined as actions that are performed on the data entered into the model, and the process may be one of the massive ready-to-use system tools available within ArcGIS, which can be easily dragged from ArcToolbox and dropped into the model implementation interface Modelbuilder window, and may be code. They are created using Python or any other programming language. The operation procedure is complex and requires many operations. It is better to organize it in a separate form and then attach it to the basic form, as it appears in the basic form in the form of a separate rectangle (Schaller and Mattos 2009). There is no relative sea level rise surface along the NDC that can be used to build a model in a GIS environment to assess the future impact on land-cover classes, roads, and population. GIS is used to predict the future impact of SLR based on the available data and global scenarios and local measurements.

The model consists of parameters and processes in four consecutive stages, as shown in Fig. 4.19, which shows the executive implementation steps of the R_{SLR} model:

First step: Algebraic multiplication (×) of the land subsidence raster layer (30 m per pixel) that contains the annual subsidence rate in (mm) in the number of years (T)

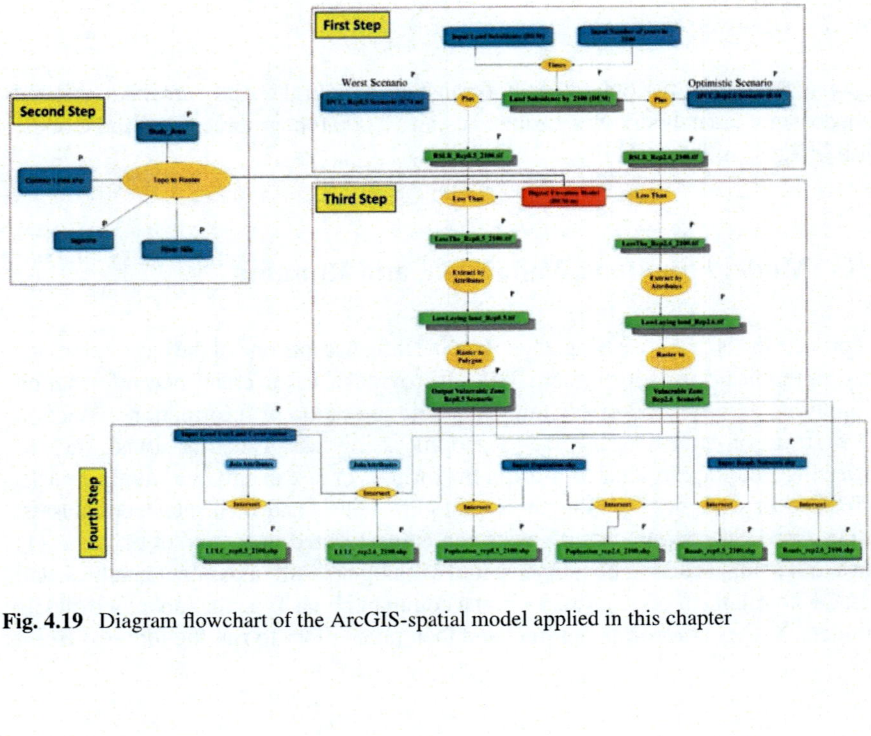

Fig. 4.19 Diagram flowchart of the ArcGIS-spatial model applied in this chapter

from the present time until the end of the twenty-first century and its value (85 years). It aims to obtain a new map of the land subsidence values along the NDC by 2100; the command (Times) of the program Trigonometric Math Function was applied.
The second step is to build a digital elevation model based on the model shown in Fig. 4.19. The SRTM (1-arc-second) data were used to build DEMs, and digital contour lines were extracted with an interval of 1 m. They were corrected and converted to a digital topographical model with an accuracy of 30-m pixels using the Topo to Raster tool (Parameter). The algebraic addition of scenario values (+) to the land subsidence layer in 2100 was used to obtain the relative sea level rise layer (R_{SLR}) in 2100. According to Eq. 4.2 and was executed using the command **PLUS** (+) of the Arc Toolbox Trigonometric Math Function commands.
Third step*:** The subtraction (−) of the relative sea level rise layer at 2100 m (Raster) from the DEM model layer for all scenarios was used to identify and separate areas at risk of sea level rise at each scenario by 2100 m. The command **LESS Than** from the Trigonometric Math Function commands was used, and the result was to obtain a single cell layer containing only two values (1) representing the land that will be below sea level, and the value of (0) represents the highest lands in the level that are outside the risk zone until 2100. The risk zone was automatically extracted using the ***Extract by Attributes command by applying the following expression (Expression: Value $\leq$ 0), followed by a conversion from Rater to Vector using the Raster to Polygon tool command. The results should be vulnerable maps according to several scenarios.
Fourth step: Input the Maps (Land-Use/Land-Cover Map) (Vector), Population density Map (Vector), and Road Network Map (Vector). This is to assess the future impacts of SLR on population and human activities along the NDC. Subtraction of the vulnerable zone was performed using the INTERSECT command within Arc Toolbox Overlay Tools.
Finally, we analyze the final vulnerability maps for each potential scenario and estimate the impact on human activities.

Before running the model, validation of all operations is needed. The variables and data must be checked to ensure that all data elements and values of variables are valid and ready-to-run. Last, the Run Entire Model command line is responsible for passing all processes in the model that are in a ready-to-apply state (Allen 2011).

4.11 Model Validation and Running

Model validation means verifying that all data elements and parameter values are valid. Validating will return has-been-run processes to their ready-to-run state. If, by validating the model, processes that were in the ready-to-run or has-been-run state change to not-ready-to-run (elements are white), then one or more input variables are invalid (ESRI 2015).[1]

[1] https://desktop.arcgis.com/en/desktop/latest/analyze/modelbuilder/validating-A-model.htm.

ESRI indicated that models can be run from within ModelBuilder or from the model tool dialog box, Python window, or in a script. When running a model within ModelBuilder, either the entire model or some selected processes can be run. When processes are running in ModelBuilder, their execution can be followed in the geoprocessing progress dialog box.

There are three ways to run a model from within ModelBuilder (Fig. 4.20):

- **Run Single tool**—Select one tool, right-click, then click Run. Earlier processes in the chain also run if needed. Later processes in the chain do not run; however, if they are in the has-been-run process state, their state changes back to ready-to-run.
- **Run Ready-to-run tools**—Click Run from the Model menu or the run tool on ModelBuilder toolbar. All tools that are in a ready-to-run state run.
- **Run entire model**—Click Run Entire Model from the Model menu. This runs all tools that are in a ready-to-run state.

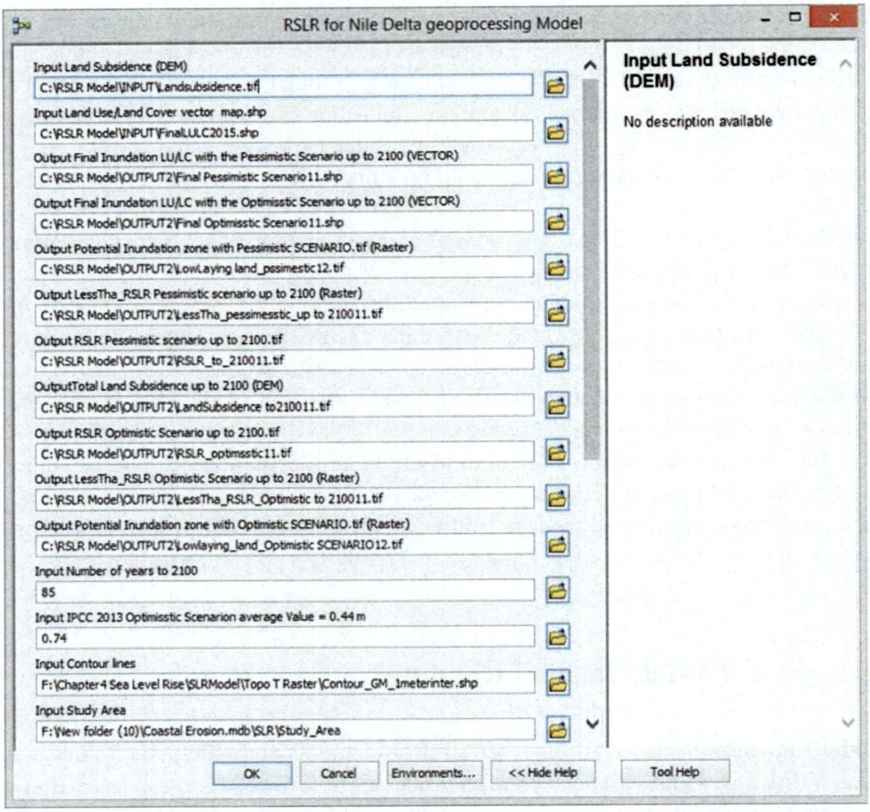

Fig. 4.20 Graphical interface of the geoprocessing model designated for use in this study to calculate the RSLR and potential impact builder environment

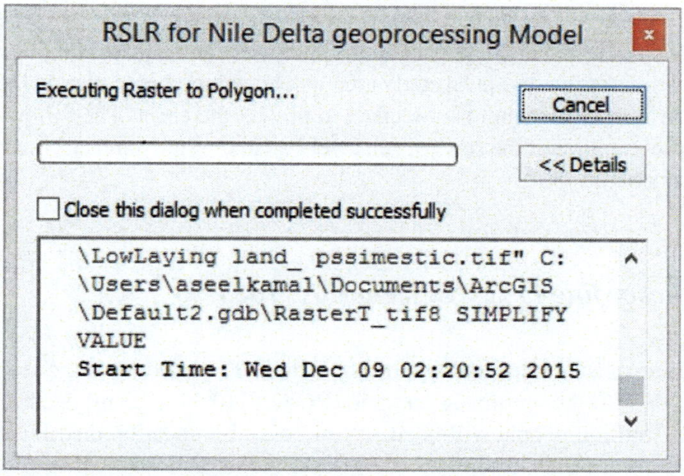

Fig. 4.21 ArcGIS model validation and running

According to ESRI (2015), when a model is run, all model variables are validated, and those tools in a ready-to-run state are executed. If any output variables should be added to display, the outputs are added to the ArcMap table of contents. Tools that are not in a ready-to-run state or are dependent on a tool that is not ready-to-run do not execute. There are some key differences when running a model from within ModelBuilder and running a model from its tool dialog box. When a model is run within Model builder, the following apply:

- Intermediate data are preserved between model runs. These data are also preserved if the model is saved, closed, and reopened for editing. When a model is run from the tool dialog box, intermediate data are deleted.
- All the outputs with a check mark next to Add to Display are added to the display.
- Progress will be shown in the geoprocessing progress dialog box, as the process is executing in the foreground. Models run within ModelBuilder never execute as a background process. Models executed from the tool dialog box can be run in the background.
- A result is not written to the Results window (Fig. 4.21).

4.12 Results and Discussion

The analysis of the output of the GIS model for both the optimistic and pessimistic scenarios indicates that there are potential hazards that may affect the coastal zone of the Nile Delta coast if no action is taken. The effect of sea level rise on the coast of the Nile Delta began to appear in the seepage of salty seawater into the fresh aquifer, as the salty water surrounded the fresh groundwater in the reservoir from all sides, and it began to leak in the south toward the land, and the salinity increased with depth (Mabrouk et al. 2013). It is clear from the spatial analysis of the possible rise in the relative sea level and its impact on the study area and human activities that the northern margins of the low-level study area will be vulnerable to marine inundation unless the necessary precautions are taken to protect the coast. The following is an analytical presentation of the risks of sea level rise in the study area according to the possible scenarios.

4.12.1 According to IPCC Scenarios (2013)

The estimation of SLR impact along the NDC by 2100 according to the scenarios of the IPCC (2013). If the optimistic scenario (RCP2.6) (0.39–0.83 m) occurs, approximately 13.7 million people will be at risk of coastal inundation distributed over 8 coastal governorates and 63 administrative departments, and the most affected governorates are Alexandria, Beheira, Kafr El Sheikh, Dakahlia, and Damietta. Under the pessimistic RCP8.5 scenario (0.69–1.13 m), 14.5 million people would be at risk of sea level rise, distributed over 8 coastal governorates and 70 administrative centers and departments, with the most affected governorates being Alexandria, Beheira, Kafr El Sheikh, Dakahlia, and Damietta.

The impact of SLR on land use/land cover, according to the best scenario RCP2.6, more than 4900 km^2 will be inundated, of which 2589 km^2 will be agricultural land and 136 km^2 will be populated urban land, and 1423 km^2 of wetlands and fish farms will be affected. Sea rise also threatens roads, corridors, bridges, and railways, and it is expected that 186.6 km of major road lengths will be in danger.

The worst scenario RCP8.5 will threaten more than 5,500 km^2 that will be at risk, approximately 2,845 km^2 of productive agricultural land will be at risk, and 149 km^2 of populated urban areas will be threatened. An area of 1,680 km^2 of fish farms and wetlands will be threatened, and 225.3 km of roads will be affected by marine submersion, which is the largest prediction of sea level rise in the study area, as shown in Tables 4.6, 4.7 and Fig. 4.23.

4.12.2 According to IPCC Scenarios (2007)

Estimation of the impact of sea level rise along the NDC by the end of the twenty-first century using the IPCC (2007) scenarios indicated that the best scenario B1 (0.33–0.77 m) might affect 12 million people's marine inundation risk. Those people were distributed over eight coastal governorates and 56 administrative centers and departments, with the governorates of Alexandria, Beheira, Kafr El Sheikh, Dakahlia, and Damietta. However, according to the worst-case scenario A1F1 (0.54–0.98 m), approximately 12.13 million people will be at risk of sea level rise by 2100, distributed over 8 coastal governorates and 59 administrative centers and departments: Alexandria, Beheira, Kafr El Sheikh, Dakahlia, and Damietta.

Table 4.6 Estimation of the potential impact of SLR along Egyptian coastal governorates under IPCC and Mediterranean tide-gauge scenarios

Governorates	IPCC(2013)				IPCC(2007)				Tide gauge	
	Optimistic scenario		Pessimistic scenario		Optimistic scenario		Pessimistic scenario			
	Rcp 2.6		Rcp 8.5		B1		A1F1			
	Population	Counties	Population	Counties	Population	Counties	Population	Counties	Population	Counties
Alexandria	3.981.237	14	3.981.237	14	3.598.464	12	3.601.018	13	3.598.464	12
Behira	3.855.610	13	4.082.417	14	3.232.688	11	3.232.688	11	3.232.688	11
Kafr El Sheikh	2.182.458	9	2.182.458	9	1.596.959	7	1.596.959	7	1.596.959	7
Dakahlia	1.511.708	7	1.793.496	9	1.511.708	7	1.511.708	7	1.511.708	7
Damietta	1.046.722	7	1.058.713	8	1.046.722	7	1.046.722	8	969.427	6
Sharqia	500.053	1	673.716	2	500.053	1	500.053	1	500.053	1
Port Said	494.523	10	570.603	12	433.177	9	494.523	10	433.177	9
Ismailia	147.463	2	147.463	2	147.463	2	147.463	2	147.463	2
Total	13.719.774	63	14.490.103	70	12.067.234	56	12.131.134	59	11.989.939	55

Table 4.7 Estimation of the potential impact of SLR human activities and populations under different IPCC scenarios

Scenarios		Klein, Lichter	IPCC (2007)		IPCC (2013)	
		(0.17–0.61 m)	B1	A1F1	Rcp 2.6	Rcp 8.5
			(0.33–0.77 m)	(0.54–0.98 m)	(0.39–0.83 m)	(0.69–1.13 m)
		By 2100	Optimistic scenario	Pessimistic scenario	Optimistic scenario	Pessimistic scenario
Land Use/ Land Cover (km^2)	Agriculture Land	2443.93	2548.28	2700.08	2589.15	2845.99
	Desert	110.58	123.09	142.41	128.13	168.24
	Urban and BuiltUp Area	127.53	133.9	142.015	136.18	149.1
	Wetland	1308.67	1390.97	1516	1423.3	1680.06
	Water	625.22	637.76	650.91	641.13	661.65
Total Risk Areas (km^2)		4615.93	4834	5151.415	4917.89	5505.04
Highway km		170.9	182.13	200.94	186.5	225.3

Table 4.6 and Fig. 4.22 illustrate the impact of sea level rise on land-use units/land cover according to the best-case scenario B1. More than 4800 km^2 will be flooded, approximately 2548 km^2 will be lost from agricultural land, and 133 km^2 of populated urban land will be under the sea. It is estimated that 1390 km^2 of wetlands and fish farms will be affected, and the main roads will lose approximately 182.13 km of the vital highway. According to the worst-case scenario A1F1, more than 5,100 km^2 will be at risk, and 2,700 km^2 of agricultural land, 142 km^2 of populated built-up areas, fish farms, and wetlands will be affected with an area of 1,516 km^2. According to the worst scenario, more than 200 km of roads will be at risk. This forecast, which has a maximum sea level altitude in the worst-case scenario (0.98 m), is close to the IPCC forecast. 2013, which has the maximum sea level altitude according to the worst-case scenario (1.13 m).

4.12.3 According to the Tide Gauge of Mediterranean

The calculation of SLR potential impacts according to the tide-gauge measurement of the eastern Mediterranean shows that, as the maximum sea level height reached (0.17–0.61 m), approximately 11.9 million people will be at risk of marine inundation by 2100. Those people are distributed over eight coastal vulnerable governorates, and 55 administrative departments, including Alexandria, Beheira, Kafr El Sheikh, and Dakahlia, are among the most vulnerable governorates. It is also clear from the analysis of Table 4.7 that a coastal zone of 4,615 km^2 will be submerged

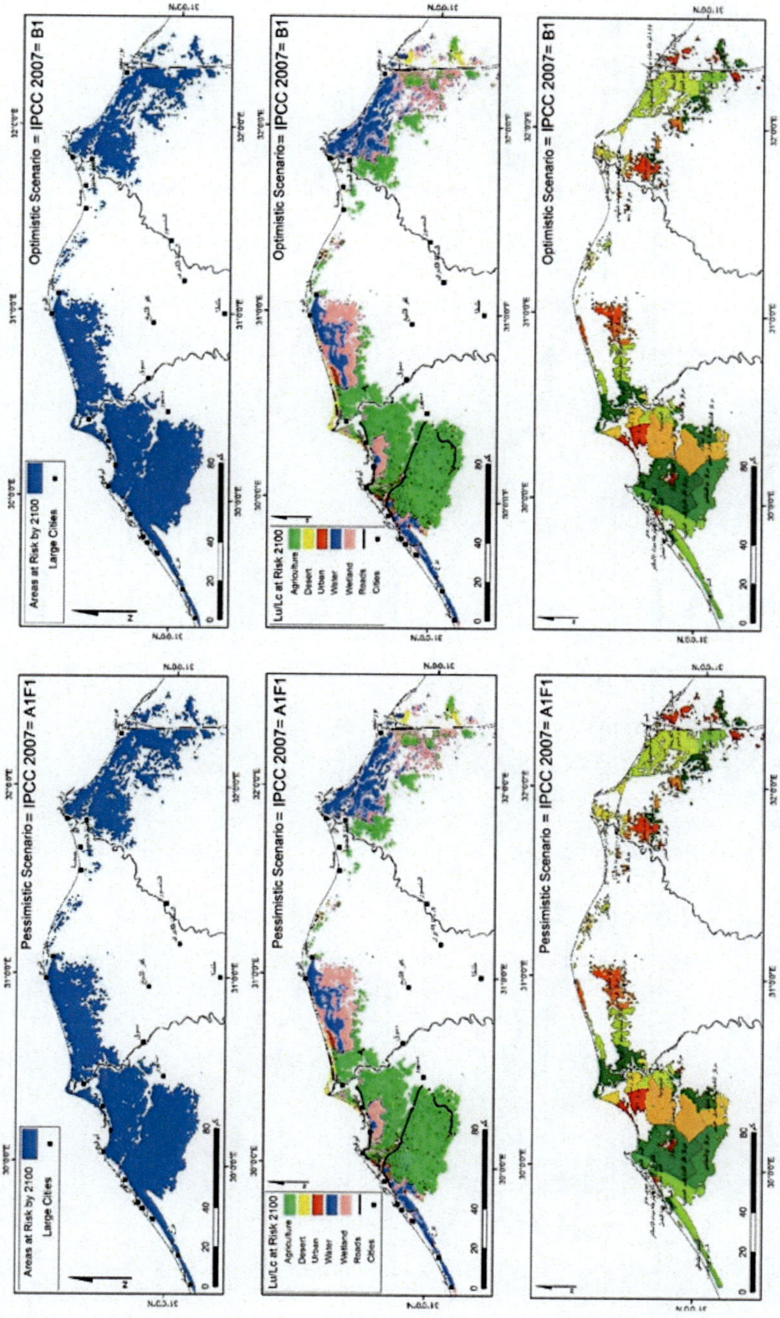

Fig. 4.22 Vulnerable Maps of R_{SLR} by 2100. **a** Optimistic Scenario according to IPCC (2007), **b** Pessimistic Scenario according to IPCC (2007)

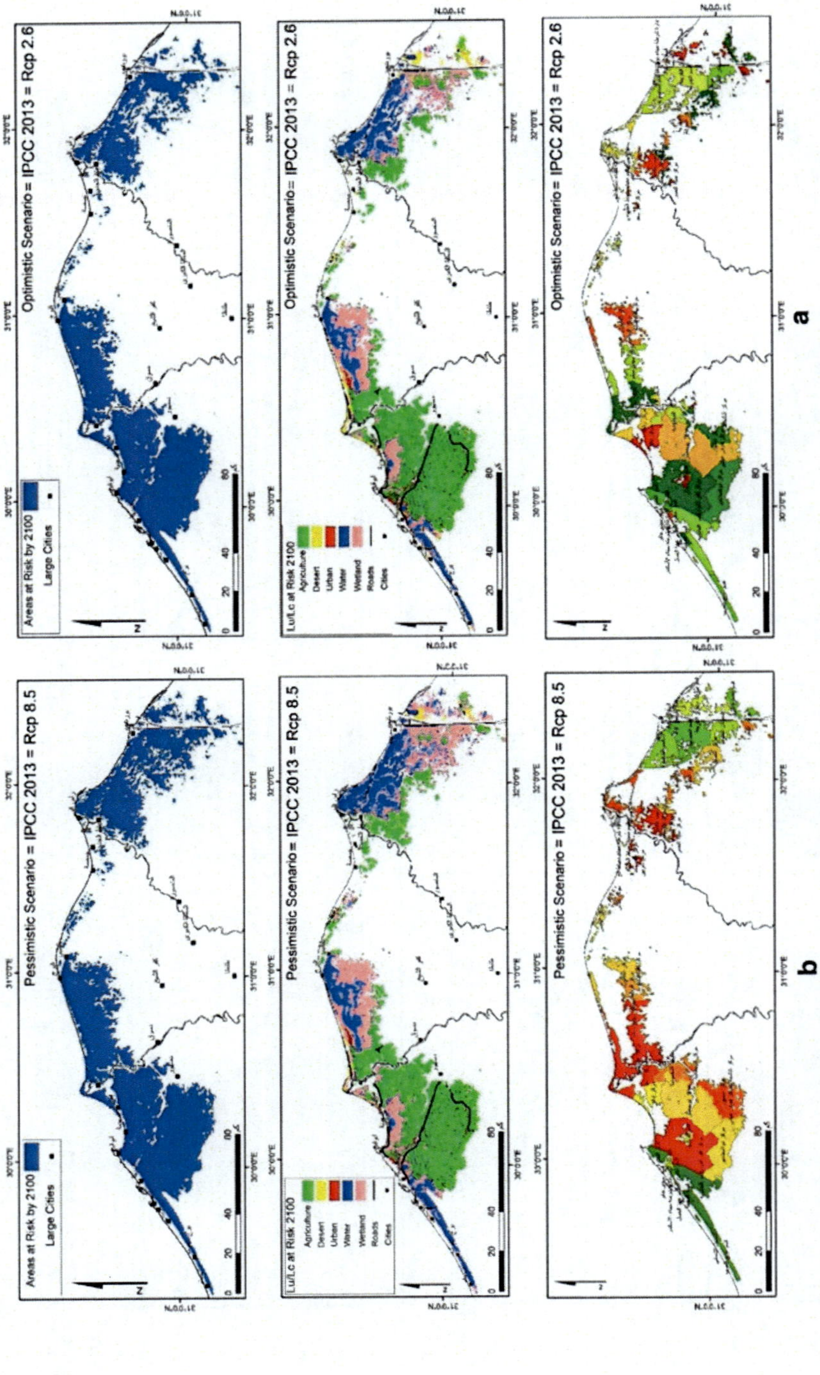

Fig. 4.23 Vulnerable Maps of R_{SLR} by 2100. **a** Optimistic Scenario according to IPCC (2013), **b** Pessimistic Scenario according to IPCC (2013)

under the sea. Approximately 2,443 km^2 of productive agricultural land, 127 km^2 of populated built-up areas, 1,308 km^2 of wetlands and fish farms, and more than 170 km of main roads, including international coastal roads, will become in the hazardous zone according to this scenario. According to the results of the Mediterranean tide measurement scenario, the maximum retreat of the shoreline by 2,100 m will be 48.6 km, 24.6 km, 50 km, and 42.8 km in the areas of Abu Qir Bay, Burullus, Manzala, and Tienah coastal plain, respectively. According to the worst-case scenario IPCC, 2013 (RCP2.8), the maximum retreat of the shoreline by 2100 will be 49.4 km, 26.2 km, 50.2 km, and 43.18 km in the areas of Abu Qir Bay, Burullus, Lake Manzala, and Tine Plain, respectively.

4.13 Summary

This chapter examined and analyzed one of the most important problems facing global low-laying coasts in the world. Sea level rise is an ongoing phenomenon, and there are several factors controlling it. SLR is not a modern problem, and many geoindicators refer to the ancient submergence of NDC during the Holocene. The objective of this chapter is to assess the potential risks of relative sea level rise on human activities and population until 2100 based on five possible global and regional scenarios presented by the World Panel on Climate Change IPCC (2007, 2013) and predication according to measurements taken from the Mediterranean tide gauge. The study applied GIS technology in spatial modeling, which depends on a set of basic variables affecting SLR (topography, land subsidence, R_{SLR}, LU/LC, population density, and road network). Spatial modeling can help to link multicriteria together using a set of tools and functions to map the future impacts of SLR by the end of the twenty-first century. The study concluded that the lowest possible scenario according to the tide-gauge scale in the Mediterranean Sea is expected to reach the maximum relative sea level rise to 0.61 m by 2100, which makes 11.9 million people at risk and threatens them to leave their homes. Those people are distributed over 8 coastal governorates and 55 centers, and administrative departments Alexandria, Beheira, Kafr El Sheikh, Dakahlia, and Damietta are among the most affected governorates. The model estimated that 4,615 km^2 will be submerged, approximately 2,443.9 km^2 will be agricultural land, 127.5 km^2 will be built-up areas, and approximately 1,308 km^2 of wetlands and fish farms will be inundated. The main roads will lose 170.9 km along the NDC. According to the worst-case scenario applied to the study area, IPCC, 2013, RCP 8.5, which predicts that the relative sea level will rise to 1.13 m by 2100. According to this scenario, 14.5 million people will be at risk, distributed spatially along eight coastal governorates and 70 administrative centers and departments. The Nile Delta will lose approximately 5,505 km^2 of its land, including 2,845 km^2 of productive agricultural land and 149 km^2 of urban built-up areas, while the wetlands and fish farms will lose 1,680 km^2 by marine inundation, and the analysis showed that more than 225 km of coastal main roads would be threatened submerged under the sea.

References

AASTMT/Egis BCEOM International (2011) Phase 1: risk assessment for the present situation and horizon 2030—Alexandria area. In: Climate change adaptation and natural disasters preparedness in the coastal cities of North Africa, 31 January 2011 (GED 80823T). Available at http://www.egis-bceominternational.com/pbm/

Allen DW (2011) Getting to know ArcGIS modelbuilder. ESRI Press

Allison I et al (2009) The Copenhagen diagnosis: updating the world on the latest climate science. Climate Change Research Centre (CCRC), the University of New South Wales, Sydney, Australia

Aly M, Andrew G, Howard J (2012) Land subsidence in the Nile Delta of Egypt observed by persistent scatterer interferometry. Remote Sens Lett 3(7): 621–630

Anderson JR, Hardy EE, Roach JT, Witmer RE (1976) A land use and land cover classification system for use with remote sensor data. Reston VA, U.S. Geological Survey Professional Paper 964, p 28

Becker R, Sultan M (2009) Land subsidence in the Nile delta: inferences from radar interferometry. Holocene J 19:949–954

Benjamin J, Bicket A, Anderson D, Hale A (2014) A multi-disciplinary approach to researching the intertidal and marine archaeology in the Outer Hebrides, Scotland. J I Coast Archaeol 9:400–424

Bird E (1996) Coastal erosion and rising sea level. In: Millimann JD, Haq BU (eds) Sea level rise and coastal subsidence. Kluwer Academic Publishers, Dordrecht, The Netherlands, pp 87–103

Brochier F, Ramieri E (2001) Climate change impact of the Mediterranean coastal zones. Fondazione Eni Enrico Mattei, Milano

Bruun P (1962) Sea level rise as a cause of shoreline erosion. J Waterw Harbors Div 88(1962):117–130

Bruun P (1988) The Bruun rule of erosion by sea level rise: a discussion of large-scale two- and three-dimensional usages. J Coast Res 4:627–648

Cabanes C, Cazenave A, Le Provost C (2001) Sea-level rise during past 40 years determined from satellite and in situ observations. Science 294:840–842

Cazenave A, Cabanes C, Dominh K, Gennero MC, Le Provost C (2003) Present day sea level change: observations and causes. Space Sci Rev 108:131–144

Cazenave A, Llovel W (2010) Contemporary sea level rise. Annu Rev Mar Sci 2(1):145–173

Cazenave A, Remy F (2011) Sea level and climate: measurements and causes of changes. Interdisc Rev Clim Change 2(5):647–662

Chang K-t (2014). Introduction to geographic information systems, 7th edn. McGraw Hill Higher Education, Boston, p 425

Church JA, Gregory JM, Huybrechts P, Kuhn M, Lambeck K, Nhuan MT, Qin D, Woodworth PL (2001) Changes in sea level. In: Houghton JT, Ding Y, Griggs DJ, Noguer M, van der Linden PJ, Dai X, Maskell K, Johnson CA (eds) Climate change 2001: the scientific basis. Contribution of working group I to the third assessment report of the intergovernmental panel on climate change. Cambridge University Press, Cambridge, pp 639–693

Church JA, Nicholls R, Hay JE, Gornitz V (2007) Ice and sea-level change. In: Earner J (ed) Global outlook for ice and snow. United Nations Environment Programme, New York

Church JA, White NJ, Coleman R, Lambeck K, Mestrovic JX (2004) Estimates of the regional distribution of sea-level rise over the 1950 to 2000 period. J Clim 17(13):2609–2625

Church JA, White NJ, Hunter JR, Lambeck K (2008) Briefing: a post-IPCC AR4 update on sea-level rise. Antarctic Climate & Ecosystems Cooperative Research Centre

Church JA, White NJ (2011) Sea-level rise from the late 19th to the early 21st century. Surv Geophys 32:585–602

Coutellier V, Stanley JD (1987) Late quaternary stratigraphy and paleogeography of the eastern Nile delta, Egypt. Mar Geol 77(3–4):257–275

Dasgupta S, Laplante B, Meisner C, Wheeler D, Yan J (2009) The impact of sea level rise on developing countries: a comparative analysis. Clim Change 93:379–388

Dean RG, Houston JR (2013) Recent sea level trends and accelerations: comparison of tide gauge and satellite results. Coastal Eng 75:4–9

Deabes EA (2003) The study of sea level changes and currents at Rosetta and Damietta outlets and Abu Quir Bay. M.Sc. Thesis, Faculty of Science, Alexandria University, pp 242

Dobosiewicz J (2001) Applications of DEM and GIS to coastal flood studies along the Shoreline of Raritan Bay, New Jersey. Environ Geosci 8(1):11–20

Douglas BC (2000) An introduction to sea-level. In: Douglas BC, Kearney MF, Leatherman SP (eds)), Sea-level rise history and consequences. Academic Press, New York, USA, pp 1–11

Douglas BC (2001) Sea level change in the era of the recording tide gauge. In: Douglas BC, Kearney MS, SP Leatherman (eds) Sea Level Rise: History and Consequences, International Geophysics Series, vol 75. Elsevier, New York, pp 37–64

Ebaid H (2014) Accuracy enhancement of SRTM and ASTER DEMS using weight estimation regression model. IJRET: Int J Res Eng Technol 3(8):371–377

Eckert S, Jelinek R, Zeug G, Krausmann E (2012) Remote sensing-based assessment of tsunami vulnerability and risk in Alexandria, Egypt. Appl Geogr 32:714e723

El Askary MA, Frihy OE (1986) Depositional phases of Rosetta and Damietta promontories on the Nile Delta coast. J Afr Earth Sci 5:627–633

El Nahry AH, Doluschitz R (2010) Climate change and its impacts on the coastal zone of the Nile Delta. Environ Earth Sci 59:1497–1506

El Raey M, Dewidar K, El Hattab M (1999) Adaptation to the impact of sea level rise in Egypt. Clim Res J 12:117–128

El-Raey M (1997) Vulnerability assessment of the coastal zone of the Nile delta of Egypt, to the impacts of sea level rise. Ocean Coast Manag Clim Change Integr Coast Manag 37(1):29–40

El-Raey M, Dewidar K, El Hattab M (1999) Adaptation to the impact of sea level rise in Egypt. J Clim Res 12:117–128

El-Raey M, Nasr S, Frihy OE, Desouk S, Dowidar K (1995a) Potential impacts of accelerated sea-level rise on Alexandria Governorate, Egypt. J Coast Res 51:190–204

El-Raey M, Nasr S, El-Hattab M, Frihy OE (1995b) Change detection of Rosetta promontory over the last forty years. Int J Rem Sens 16:825–834

El-Raey M, Fouda Y, Nasr S (1997) GIS assessment of the vulnerability of Rosetta area, Egypt to impacts of sea rise. J Environ Monit Assess 47:59–77

Emery KO, Aubrey DG, Goldsmith V (1988) Coastal neotectonics of the Mediterranean from tide-gauge records. Mar Geol 81:41–52

Ericson JP, Vorosmarty CJ, Dingman SL, Ward LG, Meybeck L (2006) Effective sea level rise and deltas: causes of change and human dimension implications. Glob Planet Change 50:63–82

ESRI (2000) ModelBuilder for ArcView spatial analyst 2. ESRI Press, Redlands, CA, p 18

ESRI (2004) ArcGIS desktop version 9. ESRI, Redlands, CA

ESRI (2015) What is ModelBuilder? https://desktop.arcgis.com/en/arcmap/latest/analyze/modelbuilder/what-ismodelbuilder.htm

Farr TG, Rosen PA, Caro E, Crippen R, Duren R, Hensley S, Kobrick M, Paller M, Rodriguez E, Roth L (2007) The shuttle radar topography mission. Rev Geophys 45.

Frihy OE (1992a) Sea-level rise and shoreline retreat of the Nile delta promontories, Egypt. Nat Hazards 5:65–81

Frihy OE (1992b) Beach response to sea level rise along the Nile delta coast of Egypt. In Woodworth PL (ed) Sea level changes: determination and effects geophysical monograph 69, vol 2. American Geophysical Union, IUGG, pp 81–85

Frihy OE (1992c) Holocene sea level changes at the Nile Delta Coastal zone of Egypt. GeoJournal 26(3):389–394

Frihy OE (2003) The Nile delta-Alexandria coast: vulnerability to sea-level rise, consequences and adaptation. Mitig Adapt Strat Glob Change 8:115–138

Frihy OE, Dewidar K, El-Raey M (1996) Evaluation of coastal problems at Alexandria, Egypt. Ocean Coast Manage 30:281–295

Frihy OE, El-Sayed M (2013) Vulnerability risk assessment and adaptation to climate change induced sea level rise along the Mediterranean coast of Egypt. Mitig Adapt Strateg Glob Change 18:1215–1237. https://doi.org/10.1007/s11027-012-9418-y

Frihy OE, Essam A, Samy M, Farid A (2010) Alexandria-Nile Delta coast, Egypt: update and future projection of relative sea-level rise. Environ Earth Sci 61:253–273. https://doi.org/10.1007/s12665-009-0340-x

Frihy OE, Hassan MS, Deabes EA, Badr AM (2008) Seasonal wave changes and the morphodynamic response of the beach-inner shelf of Abu Qir Bay, Mediterranean coast, Egypt. Mar Geol 247:145–158

Goddio F (2004) Canopus I: the submerged western Canopic region. Periplus Publishing, London

Goddio F (2007) The topography and excavation of Heracleion–Thonis an East Canopus (1996–2006). Oxford Centre for Maritime Archaeology, Monograph 1, Institute of Archaeology, Oxford, p 136

Goddio F (1998) Alexandria the submerged royal quarters. Periplus, London, p 274

Goodchild MF (2005) GIS, Spatial analysis and modeling overview. In: Maguire D, Batty M, Goodchild MF (eds) GIS, spatial analysis, and modeling. ESRI Press, Redlands, CA, pp 1–18

Goodchild MF (1992) Geographical information science. Int J Geogr Inf Syst 6(1):31–45

Gong P, Wang J, Yu L, Zhao YC, Zhao YY, Liang L, Niu ZG, Huang XM, Fu HH, Liu S (2013) Finer resolution observation and monitoring of global land cover: first mapping results with Landsat TM and ETM+ data. Int J Remote Sens 34:2607–2654

Grinsted A, Moore JC, Jevrejeva S (2009) Reconstructing sea level from paleo and projected temperatures 200 to 2100 AD. Clim Dyn. https://doi.org/10.1007/s00382-008-0507-2

Gumma MK, Thenkabail P, Hideto F, Nelson A, Dheeravath V, DawuniBusia D, Rala A (2011) Mapping irrigated areas of Ghana using fusion of 30 m and 250 m resolution remote-sensing data. Remote Sens 3:816–835

Gutierrez BT, Plant NG, Pendleton EA, Thieler ER (2014) Using a Bayesian Network to predict shoreline change vulnerability to sea-level rise for the coasts of the United States: U.S. Geological Survey Open-File Report 2014–1083, p 26

Hasan E, Khan SI, Hong Y (2015) Investigation of potential sea level rise impact on the Nile Delta, Egypt using digital elevation models. Environ Monitor Assess 187(10):649. https://doi.org/10.1007/s10661-015-4868-9

Hassaan MA (2013) GIS-based risk assessment for the Nile Delta coastal zone under different sea level rise scenarios case study: Kafr El Sheikh Governorate. Egypt J Coast Conserv 17:743–754. https://doi.org/10.1007/s11852-013-0273-0

Hassaan MA, Abdrabo MA (2013) Vulnerability of the Nile Delta coastal areas to inundation by sea level rise. Environ Monitor Assess 185(8):6607–6616. https://doi.org/10.1007/s10661-012-3050-x

Hereher M (2010) Vulnerability of the Nile Delta to sea level rise: an assessment using remote sensing. Geomatics, Nat Hazards Risk 1(4):315–321

Hesse A (1998) Arguments pour une nouvelle hypothèse de localization de l'Heptastade d'Alexandrie. Alexandrina 1:1–33 [in French]

Horton RC, Herweijer C, Rosenzweig J, Gornitz V, Ruane AC (2008) Sea level rise projections for current generation CGCMs based on the semiempirical method. Geophys Res J Letter 35:L02715. https://doi.org/10.1029/2007GL032486

IPCC (1998) North America. In: Watson RT, Zinyowera MC (eds) The regional impacts of climate change: an assessment of vulnerability. Special report of international panel on climate change, working group II. Cambridge University Press, Cambridge, pp 253–330

IPCC (2007) Climate change 2007: impacts, adaptation, and vulnerability. In: Parry ML, Canziani OF, Palutikof JP, van der Linden PJ, Hanson CE (eds) Contribution of working group II to the fourth assessment report of the intergovernmental panel on climate change. Cambridge University Press, Cambridge, UK

IPCC (2013) Climate change 2013: the physical science basis. In: Contribution of working group I to the fifth assessment report of the intergovernmental panel on climate change. Cambridge University Press, Cambridge, UK and New York, p 1535

Irons JR, Dwyer JL, Barsi JA (2012) The next Landsat satellite: the Landsat data continuity mission. Remote Sens Environ 122:11–21

Iskander M (2013) Wave climate and coastal structures in the Nile delta coast of Egypt. Emirates J Eng Res 18(1):43–57

Jacob T, Wahr J, Tad Pfeffer W, Swenson S (2012) Recent contributions of glaciers and ice caps to sea level rise. Nature 482:514–518

Jensen JR (2015) Introductory digital image processing: a remote sensing perspective, 4thedn (Pearson Series in Geographic Information Science). Prentice-Hall, Englewood Cliffs.

Jevrejeva S, Grinsted MJC, Matthews A, Spada AP (2014) Trends and acceleration in global and regional sea levels since 1807. Global Planet Change 113:11–22

Klein M, Lichter M (2009) Statistical analysis of recent Mediterranean sea-level data. Geomorphology 107:3–9

Kun J, Xiangqin W, Xingfa G, Yunjun Y, Xianhong X, Bin L (2014) Land cover classification using Landsat 8 Operational Land Imager data in Beijing, China. Geocarto Int 29(8):941–951

Lo CP, Yeung AK (2007) Concepts and techniques of geographic information systems, 2nd edn. Pearson Prentice Hall, Upper Saddle River

Mabrouk MB, Jonoski A, Solomatine D, Uhlenbrook S (2013) A review of seawater intrusion in the Nile Delta groundwater system—the basis for assessing impacts due to climate changes and water resources development. Hydrol Earth Syst Sci Discuss 10:10873–10911

Manandhar R, Odeh I, Ancev T (2009) Improving the accuracy of land use and land cover classification of Landsat data using postclassification enhancement. Remote Sens 1:330–344

Meehl G et al (2007) Global climate projections. In: Climate change 2007: the physical science basis. In: Contribution of working group I to the fourth assessment report of the intergovernmental panel on climate change. Cambridge University Press, Cambridge

Miller L, Douglas BC (2004) Mass and volume contributions to twentieth-century global sea-level rise. Nature 428:406–409

Milliman D, Broadus J, Frank G (1989) Environmental and economic implications of rising sea level and subsiding deltas: the Nile and Bengal examples. Ambio 18(6):340–345

Mimura N (2013) Sea-level rise caused by climate change and its implications for society. Proc Jpn Acad Ser B Phys Biol Sci 89(7):281–301

Misdorp R, Sestini G (1976) Topography of the continental shelf of the Nile Delta. In: Proč Sem. on Nile Delta Sed. UNDP/UNESCO, Alexandria, pp 145–161

Mohanty PC, Mahendra RS, Nayak RK, Kumar TS (2017) Impact of sea level rise and coastal slope onshoreline change along the Indian coast. Nat Hazard. Published online: 23 August 2017. https://doi.org/10.1007/s11069-017-3018-9

Nafaa MG (1995) Wave climate along the Nile Delta coast. J Coast Res 11:219–229

Nicholls RJ, Cazenave A (2010) Sea-level rise and its impact on coastal zones. Science 328:1517–1520

Nicholls RJ, Wong PP, Burkett V (2007) Coastal systems and low-lying areas. In: Parry ML, Canziani OF, Palutikof JP, van der Linden P, Hanson CE (eds) Climate change 2007: impacts, adaptation and vulnerability. Contribution of working Group ii to the fourth assessment report of the intergovernmental panel on climate change. Cambridge University Press, UK

Pahlevan N, Schott JR (2013) Leveraging EO-1 to evaluate capability of new generation of Landsat sensors for coastal/inland water studies. Sel Top Appl Earth Obs Remote Sens IEEE J 6(2)

Papadopoulos GA, Fokaefs A (2005) Strong tsunamis in the Mediterranean Sea: a re-evaluation. ISET J Earthq Technol 42(4):159e170

Permanent Service for Mean Sea Level (2013) Alexandria tide gauge data. Available http://www.psmsl.org/data/obtaining/stations/503.php

Prasad S (2016) Remotely sensed data characterization, classification, and accuracies; chapter processing remote-sensing data in cloud computing environments. CRC Press, Boca Raton, FL, USA

Rabus B, Eineder M, Roth A, Bamler R (2003) The shuttle radar topographic mission—a new class of digital elevation models acquired by spaceborne radar. ISPRS J Photogramm Remote Sens 57:241–262.

Said R (1981) The geological evolution of the river Nile. Springer-Verlag, New York, p 151

Schaller J, Mattos C (2009) GIS model applications for sustainable development and environmental planning at the regional level. In: GeoSpatial Visual Analytics, pp 45–57

Sestini G (1989) Nile Delta: a review of depositional environments and geological history. In: Whateley MKG, Pickering KT (eds) Deltas: sites and traps for fossil fuels. Geological Society Special Publication 41:99–127

Shepard FP (1973) Submarine geology, 3rd edn. Harper and Row, New York, p 517

Smits PC, Dellepiane SG, Schowengerdt RA (1999) Quality assessment of image classification algorithms for land-cover mapping: a review and a proposal for a cost-based approach. Int J Remote Sens 20:1461–1486

Soliman A, El-Sharnouby B, Elkamhawy H (2014) Shoreline changes due to construction of Alexandria submerged breakwater, Egypt. In: Lehfeldt R, Kopmann R (eds) ICHE 2014. Bundesanstalt für Wasserbau, Hamburg. ISBN 978-3-939230-32-8

Stanley DJ (1990) Recent subsidence and northeast tilting of the Nile delta, Egypt. J Mar Geol 94:147–154

Stanley DJ (1997) Mediterranean deltas: subsidence as a major control of relative sea-level rise. Bulletin de I'lnslilul occanographique, Monaco, n special 18, ClESM Science Series n 3, pp 35–62

Stanley DJ, Goodfriend GA (1997) Recent subsidence of the northern Suez canal. Sci Corresp 388:335–336

Stanley DJ, Warne AG (1993) Nile delta: recent geological evolution and human impacts. Sci J 260:628–634

Stanley DJ, Warne AG (1998) Nile delta in its destruction phase. J Coast Res 14:794–825

Stanley JD, Bernasconi MP, Jorstad TF (2008) Pelusium, an ancient port fortress on Egypt's Nile Delta coast: its evolving environmental setting from foundation to demise. J Coast Res 24(2):451–462

Stanley JD, Carlson RW, Van Beek G, Jorstad TF, Landau EA (2007) Alexandria, Egypt, before Alexander the Great: a multidisciplinary approach yields rich discoveries. GSA Today 17:4–10

Stanley JD, Jorstad TF, Goddio F (2006) Human impact on sediment mass movement and submergence of ancient sites in the two harbors of Alexandria, Egypt. Norwegian J Geology 86:337–350

Stanley JD, Toscano MA (2009) Ancient archaeological sites buried and submerged along Egypt's Nile delta coast: gauges of Holocene delta margin subsidence. J Coast Res 25(1):158–170

Stanley JD, Warne AG, Schnepp G (2004) Geoarchaeological interpretation of the Canopic, largest of the relict Nile Delta distributaries, Egypt. J Coast Res 20(3):920–930

Summerhayes CP, Sestini G, Misdorp R, Marks N (1978) Nile Delta: nature and evolution of continental shelf sediments. Mar Geol 27(1–2):43–65

Williams SJ (2013) Sea-level rise implications for coastal regions. J Coast Res (Special issue 63—understanding and predicting change in the coastal ecosystems of the northern Gulf of Mexico):184–196

Willis JK, Chambers DP, Kuo CY, Shum CK (2010) Global sea level rise, recent progress, and challenges for the decade to come. Oceanography 23:26–35

Zaid S, Mamoun M, Al-Mobark N (2014) Vulnerability assessment of the impact of sea level rise and land subsidence on north Nile Delta region. World Appl Sci J 32(3):325

Chapter 5
Geospatial Assessment and Mapping of Coastal Erosion Vulnerability Along the Nile Delta Coast: A Remote Sensing and GIS Approach

Abstract Coastal erosion is a global problem that threatens the world's low-laying coasts and is hoping under different natural and anthropogenic factors. The NDC is considered one of the most vulnerable coasts in Africa, and coastal erosion has accelerated along the NDC during the last 50 years due to the impacts of river damming and coastal protection engineering works. Globally, geospatial technology has contributed to accurately mapping and assessing coastal and marine hazards because of the large amount of remote sensing data and computational tools. The objectives of this chapter are to use remote sensing time series imagery and GIS techniques to change the analysis and mapping of coastal erosion vulnerability since 1945. Coastal erosion is critical; it can threaten the Nile Delta promontories, drain outlets, archeological and heritage sites, beaches and resorts, settlements, coastal lagoon barriers, roads and airports, and sustainable development projects. In this chapter, a detailed analysis and mapping of coastal erosion/siltation problems as well as the coastal vulnerability index were applied along the NDC.

5.1 General

Coastal erosion is a natural process that is generally only of concern when threatening human habitation or development. The main objectives of this chapter are to map and calculate the quantities of coastal processes, including erosion/accretion, along the NDC based on change detection of multitemporal satellite remote sensing imagery. It is divided into three main sections. The first section analyzes and maps the hazard of coastal erosion and sedimentation processes, including (1) analyzing the hazard of coastal erosion and shoreline retreat on the Nile River promontories, beaches and summer resorts, drain banks, lagoon barriers, and beaches of coastal cities. (2) Assessing the hazard of sedimentation processes on human activities along the navigation of the mouth of the Nile River, navigation channels of ports, Suez Canal Navigation Channels, and Lagoons Inlets. The second section examines the evaluation of the effectiveness of coastal protection means and the future situation in some places highly affected by coastal erosion, the most important of which are

K. S. Darwish, *Hazard Modeling and Assessment of the Nile Delta Coast*,
https://doi.org/10.1007/978-3-031-44324-4_5

the Nile promontories, Burullus headland, and coastal cities. The third section aims to create and analyze the coastal vulnerability map according to the global coastal vulnerability index (CVI) that was first introduced by Gornitz et al. (1994). The hazard degree zones were calculated based on eleven significant variables (geomorphological units, geological formations and deposits, topographic units, beach slope degrees, relative sea level rise [mm/yr.], shoreline change rate [m/yr.], land subsidence [mm/yr.], coastal protection work, wave height, tidal range, and population density). A GIS-based geodatabase containing all variables was created and mathematically linked using ArcGIS software, with the aim of calculating risk scores according to specific international standards that change from one coastal region to another. ArcGIS software is used to calculate the CVI criteria throughout each grid to map the digital coastal hazard map based on GIS-based mathematical computational techniques.

5.2 Coastal Erosion Hazard and Risk Along the NDC

Egypt has suffered from severe coastal erosion hazards along selected coastal areas. The NDC is one of the coasts in Egypt vulnerable to coastal deformation during the Holocene. Coastal processes are the actions of natural forces on coastal areas, which modify the shorelines and coastal landform morphology. Both natural and anthropogenic activities lead to coastal degradation and various hazards, such as shoreline changes, global warming, climate change, sea level rise, seawater intrusion, coastal erosion, floods, and tsunamis. Coastal hazard management is crucial for sustainable development, coastal management, and rehabilitation. All coastal hazards are related to geologic processes and have a profound impact on coastal landforms (Ferraras and Pararas-Carayannis 1988). The processes of coastal erosion were studied along some selected parts along the NDC. The Nile Delta is a typical wave-dominated or perhaps more accurately termed wave- and current-dominated coast. The coastal sediments along the Nile delta coast are transported literally in five different littoral subcells (Alamien Cell, Rosetta Promontory, Brullus Headland, Damietta Promontory, and Port Said subcell) (Frihy et al. 2003).

5.2.1 Erosion Risk Assessment

Changing coastline position due to erosion and sedimentation is one of the most important concerns in coastal zone management because coastlines are a very dynamic form of surface lines and can cause severe hazards to human use and development (White and Asmar 1999). Mapping and calculating their quantifications and movement trends are fundamental to safe navigation, resource management, protection, planning, and sustainable coastal environmental development (Di et al. 2003). Coastal erosion and siltation problems along the NDC are considered one of the most important environmental problems facing the Nile Delta coast, especially after the start of construction of the High Dam in 1964. Coastal erosion and associated processes also pose a threat to many human activities and means of protecting them in the study area. Many previous studies have pointed to the dangers of coastal erosion in the study area during the second half of the twentieth century and their impact on the estuaries of the Nile River and the sinks, navigational corridors of ports, barriers of coastal lakes and their gates, beaches of coastal cities, archaeological and historical sites and roads. This study aims to analyze the dangers of coastal erosion on the current two estuaries of the Nile River (Rosetta and Damietta), the beaches of resorts, the beaches of coastal villages and cities, coastal lake barriers, estuaries, and coastal roads. To achieve the objective of this study, the method of monitoring coastal changes was used based on historical topographic maps of different scales, digital satellite visuals of the Landsat type, and high-resolution satellite images published by Google Earth Pro to analyze and evaluate the dangers of erosion and coastal sedimentation and evaluate coastal protection works.

5.2.1.1 Nile Delta Promontories

There are two promontories along the NDC; the eastern promontory is located near the city of Damietta Promontory, while the western promontory is located near the city of Rosetta Promontory. These estuaries are located at the end of the Nile River journey and exceed 6000 km in length. The phenomenon of coastal erosion began at the mouths of the Nile Delta in early 1900, after the construction of the Aswan Low Dam, and the problem increased after the construction of the High Dam in 1964 (Sestini 1992).

Rosetta Promontory

The geomorphic evolution of the Rosetta Promontory during the period 1500–1900 indicates that the eastern and western flanks of the promontory extended seaward up to 11 and 8.5 km, respectively (Fanos et al. 1995). The reversal from the accretionary phase to the erosion phase occurred during the twentieth century due to the construction of delta barrages and other constructions, such as dams and barrages, which controlled the upcoming sediments. It is obviously seen that the Rosetta promontory retreated a distance of approximately 5 km from 1900 until the present. Previous studies, including UNESCO/UNDP (1978), Frihy et al. (1991), Fanos et al. (1991), Chen et al. (1992), Frihy et al. (2003), Frihy and Lawrence (2004), Hereher (2011), and Darwish et al. (2017), have classified the Rosetta promontory as one of the most vulnerable zones along the NDC.

To assess the geomorphic deformation of Nile delta promontories, a GIS-based vector change analysis of coastal changes through matching of topographic maps scale 1:25,000 produced by the Egyptian General Survey Authority in 1945 with Landsat satellite images taken in 1972, 1984, 2001, and 2015 that:

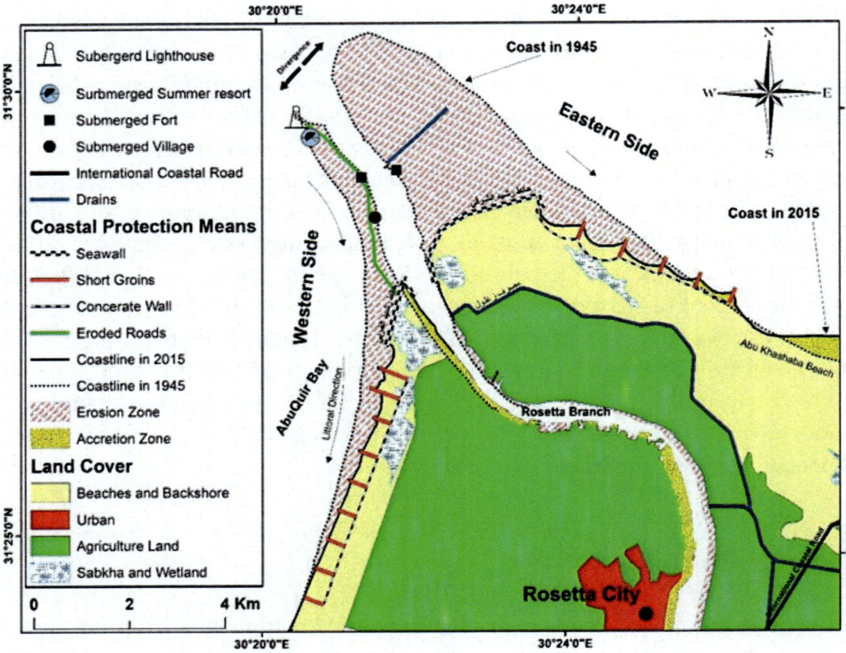

Fig. 5.1 Deformation of the Rosetta Promontory between 1945 and 2015

- The Rosetta coastline suffered the loss of land due to coastal erosion on the mouth and downstream during the period from 1945 to 2015, amounting to 17.08 km^2 (12.79 km^2 on the eastern side, 4.29 km^2 on the western side), as shown in Fig. 5.1.
- The coastline retreated to distances of 4.4 and 4.6 km on the eastern side and the western side of the promontory, respectively, during the period between 1945 and 2015. The eroded area had affected the human activities and historical sites that were completely lost and disappeared, such as the forts of Albugaz, which were built in 1870, and the Rosetta Lighthouse, which was constructed in 1942 (Frihy et al. 2004). In addition, human activities were sunken, such as roads with a length of 4.3 km and villages such as Ezbet Sidi Mansour and Oueiss Drain. The real change in Rosetta promontory and destruction of its historical activities and archeological sites are shown in Fig. 5.1 and photographed in Fig. 5.2.
- Spatial analysis of areal changes along the Rosetta promontory indicates that the promontory area decreased from 115.79 km^2 in 1972 to 100.49 km^2 in 2015, where the promontory lost −7.6 km^2 after the construction of the Aswan High Dam during (1972–1984). It also lost approximately −5.24 km^2 during the period of coastal protection construction between 1984 and 2001, and after the completion of the hard structure coastal protection engineering means to protect the promontory, the eroded area decreased to −2.45 km^2 during the period 2001–2015, as shown in Table 5.1 and Figs. 5.1 and 5.3.

Fig. 5.2 Photograph shows the impact of coastal erosion on an archaeological fort along AbuQuir Bay (taken in 2013)

Table 5.1 Deformation of Nile delta Promontories due to coastal erosion

Date	Area (sq.km)		Land loss (sq.km)		Erosion rate (sq.km/yr.)	
	Rosetta	Damietta	Rosetta	Damietta	Rosetta	Damietta
1945	113.97	187.51				
1972	115.79	187.78	+1.82	+0.26	+0.07	+0.01
1984	108.18	184.61	−7.61	−3.17	−0.63	−0.26
2001	102.94	180.94	−5.24	−3.67	−0.31	−0.22
2015	100.49	179.78	−2.45	−1.16	−0.18	−0.08

The Damietta Promontory

The general erosion and accretion patterns of this promontory are similar to those of the Rosetta promontory, where the highest rate of erosion is at the tip of the promontory, while deposition occurs on both sides of the flanks, especially the eastern side. The western side of the promontory at Ras El Bar beach experiences less variation as a result of coastal protection dating back to 1941 (Frihy et al. 2004).

Coastal change monitoring along the Damietta promontory due to erosion and sedimentation during the period from 1945 to 2015 shows that the eastern side of the Damietta promontory lost a total area of -20.84 km^2 because of the severe coastal erosion processes in front of the promontory, while an area of $+14.81$ km^2 was deposited in the form of a sand spit along the eastern flank of the promontory downward. Sand Spit extends in the southeast direction of the promontory, and therefore, the net loss of land due to marine extrusion on the eastern side of the Damietta promontory amounted to 6.03 km^2 from the land of the backshore, as shown in Fig. 5.4 and Table 5.2. The coastal erosion on the eastern side of the Damietta promontory caused destruction of the old lighthouse and the surrounding walls, and the reconstruction of a newer lighthouse was constructed approximately 220 m south of the destroyed old lighthouse. The coastal zone is protected against

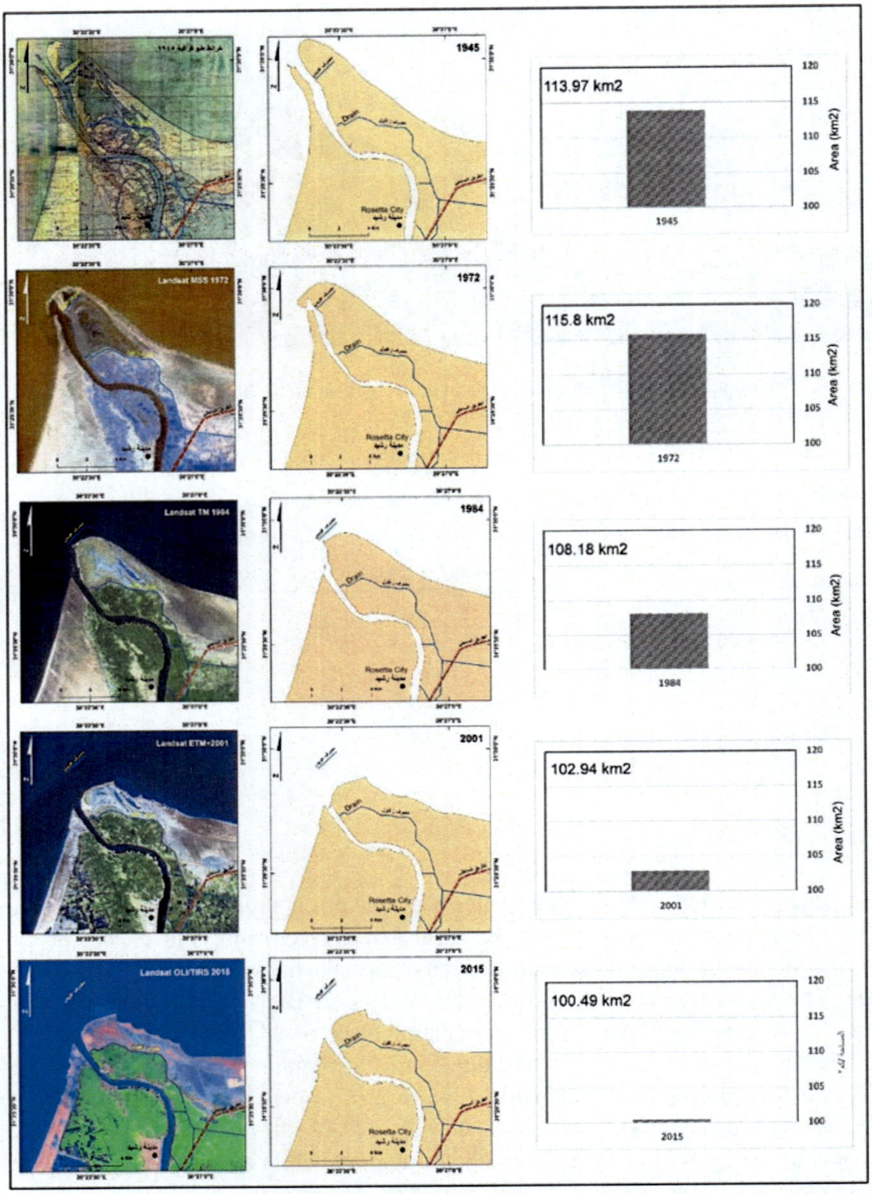

Fig. 5.3 Evolution and spatial changes of the Rosetta Promontory since 1945

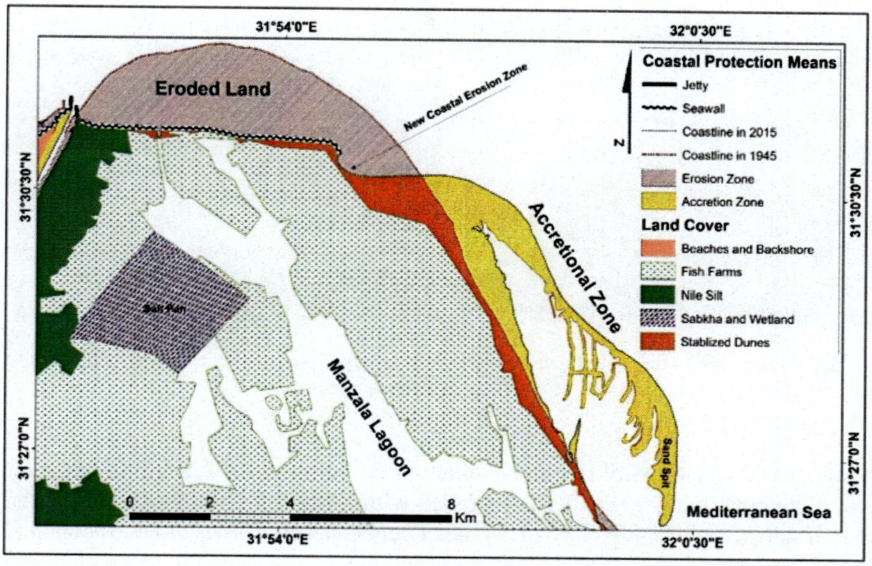

Fig. 5.4 Deformation and geomorphic evolution of the Damietta Promontory since 1945

Table 5.2 Annual coastal erosion and accretion rates along the Damietta Promontory

Period	Erosion		Accretion		Net Activity (Sq.km)
	Area (Sq.km)	Annual Rate (Sq.km/yr.)	Area (Sq.km)	Annual Rate (Sq.km/yr.)	
1945–1972	−5.63	−0.21	5.26	0.19	−0.37
1972–1984	−4.59	−0.38	2.89	0.24	−1.7
1984–2001	−5.92	−0.35	2.91	0.17	−3.01
2001–2015	−4.7	−0.34	3.76	0.27	−0.94
Total	−20.84	−0.3	14.81	0.21	−6.03

coastal erosion risk and coastline deformation, and a number of coastal engineering protection works have been constructed since 1941. A concrete seawall with a length of a few kilometers was built in 1971 to protect the coastal road between Damietta and Port Said, and most of the concrete seawall has been eroded due to the retreat of the shoreline, with only 800 m remaining on its western side until now (Frihy et al. 1991). To protect the coast of the Damietta mouth, the entrance to the Damietta channel was installed by two long jetties with lengths of 240 and 290 m on the western and eastern sides, respectively.

Along the western side of the Damietta mouth (Ras El Bar summer resort), there are many coastal protection means, such as short groins, jetties, and detached breakwaters of very hard basalt rocks Riprap, and eight detached breakwaters were installed

Fig. 5.5 Photograph of the destructive Damietta promontory. **a** Destruction of historical lighthouse along the eastern side of Damietta promontory, **b** Damietta Seawall

in the area between Ras El Bar and Damietta Port between 1991 and 2002. In addition, a concrete block seawall was installed with a length of six kilometers on the eastern side of the promontory with a load of between 4 and 7 tons of Dolos concrete blocks, which is similar to the Rosetta promontory. The analysis of erosion and sedimentation areas before and after construction showed that new erosion zones appeared on the eastern margin of the concrete wall, and the area lost due to erosion after the construction of the wall reached −4.70 km^2 at an annual rate of −0.34 km^2, as shown in Table 5.2.

Coastal erosion caused a continuous decrease in the area of the promontory zone, as shown in Fig. 5.5. Spatial analysis shows a decrease from 187.78 km^2 in 1972 to 179.78 km^2 in 2015 with a net difference of −7.73 km^2, while during the period from 2001 to 2015 after the construction of the seawall, the promontory lost an area of −1.16 km^2, as shown in Fig. 5.6.

5.2.1.2 Summer Beaches

Coastal erosion has negatively affected the beaches that are most attractive marine resorts for residents from various regions in Egypt for summer vacations during the summer months. Baltim and Ras El Bar Beaches are among the most populated and vulnerable beaches that are exposed to coastal erosion hazards and risks along the NDC.

Baltim Sea Resort

Baltim Beach extends from Burullus Inlet in the west to Al Gharbia drain Inlet in the east for up to 20 km. A comparison of spatiotemporal satellite images for successive periods (1972–1984, 1984–2001, 2001–2015) shows the following:

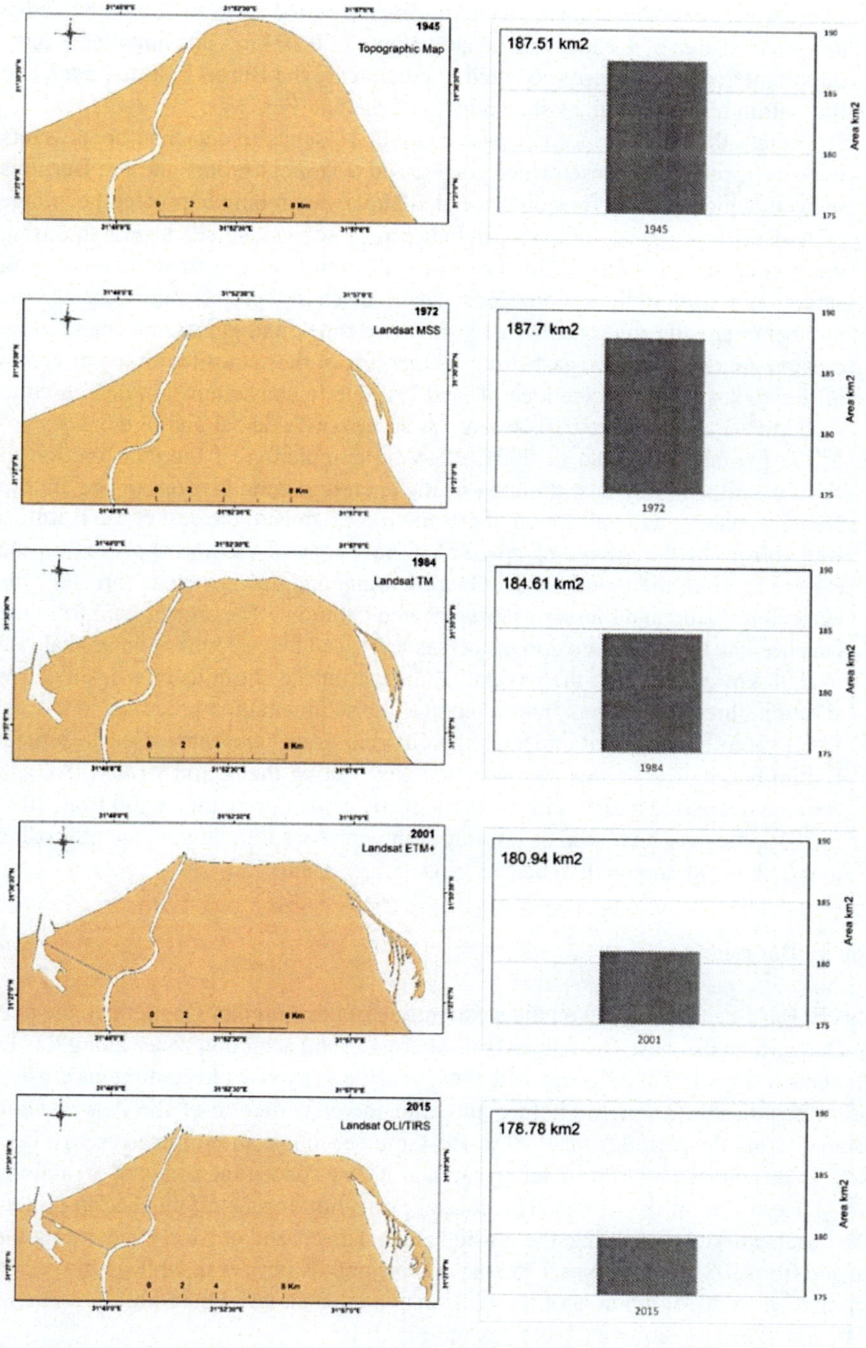

Fig. 5.6 Evolution and spatial changes of the Damietta Promontory since 1945

- There were five coastal erosion zones during the period from 1972 to 1984 with a total eroded area of 1.2 km^2 and an annual rate of 0.10 km^2, and among the most important zones that were subjected to erosion are the Burge-Baltim stretch and the Baltim resort stretch, as shown in Table 5.3 and Fig. 5.7.
- To protect the beaches from coastal erosion, 14 separate detached breakwaters were constructed in consequence stages, and 9 vertical groins and the Burullus seawall, which extends for a distance of 1.2 km, were rebuilt to protect the village of Al-Burj from coastal erosion, inundation by waves and sea level rise during the period from 1984 to 2001. There was a reversal of the erosion zones to an accretion in front of the Baltim resort due to the coastal protection works, and the number of erosion zones decreased to only two zones, and a clear movement in the erosion zones toward the east from the margins of the seawall with the direction of the downstream and the area of land lost due to coastal erosion decreased to 0.96 km^2 at an annual rate (0.06 km^2) as shown in Table 5.3 and Fig. 5.8.
- In the period from 1984 to 2001, after the completion of the construction of coastal protection means, the area of the accretion zone in front of the Baltim resort increased, and the stretch of erosion moved toward the east of the Burullus wall infront of the village of Al-Burj. A new zone of erosion appeared on the eastern drain of the mouth of the Gharbia drain, and this new zone threatens the agricultural land and tourism resorts located behind it. The area eroded from the beaches due to coastal erosion processes increased to 1.10 km^2 at an annual rate of 0.08 km^2. To protect the navigational canal of the Burullus inlet from active siltation, three jetties were built to control the sedimentation processes.
- Short groin barriers that ended at the Gharbia drain inlet were used to protect Baltim beach. The eroded area was 0.56 km^2 during the period 1972–1984, and the area decreased to 0.51 km^2 in 1984–2001, while during the period from 2001 to 2015, the area increased as a result of the presence of a large range of erosion to 0.67 km^2, as shown in Table 5.3, Figs. 5.7, 5.8, and 5.9.

Ras El Bar Beach

Ras El Bar Beach is located on the western side of the mouth of Damietta to the port of Damietta in the west. Spatial analysis of erosion and accretion zones along Ras El Bar beach since 1972–2015 shows that the beach was exposed to continuous erosion processes during the period before and after the construction of the Aswan High Dam. During the period from 1972 to 1984, the beach lost an estimated area of 0.21 km^2 at an annual rate of 0.02 km^2/year, and it was concentrated in one area along the western side of the mouth. The coastal protection means established to protect the beach worked to stabilize the erosion zone. The coastal erosion zone during the period from 1984 to 2001 was 0.75 km^2 at a rate of 0.04 km^2/year, while in the period from 2001 to 2015, the area of the erosion zone decreased to 0.34 km^2 at a rate of 0.02 km^2/year, as shown in Table 5.3 and Fig. 5.10.

Table 5.3 Spatiotemporal changes in erosion zones along Baltim and Ras El Bar Beaches

	Brullus-Baltim Shore			Ras El Bar Shore		
	1972–1984	1984–2001	2001–2015	1972–1984	1984–2001	2001–2015
	Pre-Protection	During-Protection	Post-Protection	Pre-Protection	During-Protection	Post-Protection
No. of Erosion Zones	5	2	2	1	1	1
Erosion area (sq.km)	1.21	0.96	1.1	0.21	0.75	0.34
Annual Erosion Rate (sq.km)	0.1	0.06	0.08	0.02	0.04	0.02

Fig. 5.7 Beach erosion along Baltim resort. **a** Gharbia Drain Flank, **b** Brullus Black sand zone

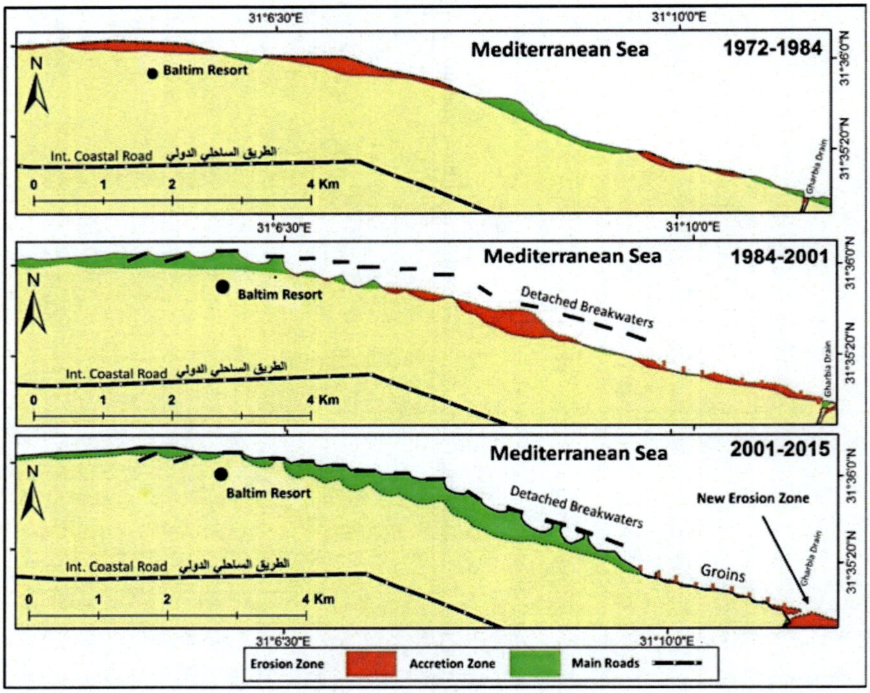

Fig. 5.8 Spatiotemporal changes in coastal erosion/accretion zones along the Baltim resort

Port Said Beach

Port Said beach extends from the entrance of the Suez Canal to El Gameil Inlet in the west, and this beach is characterized as an accretional beach. Spatiotemporal analysis of coastal erosion and accretion along the Port Said shore showed that the eroded area during 1984–2001 was 0.05 km^2, while during the period from 2001 to 2015, the beach lost an area of 0.15 km^2.

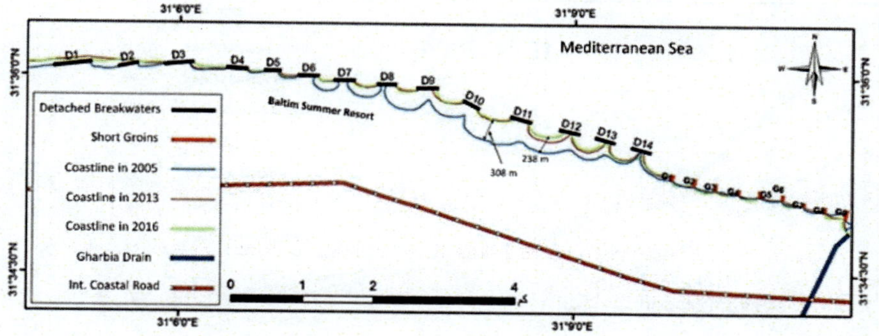

Fig. 5.9 Coastal protection means along the Baltim resort

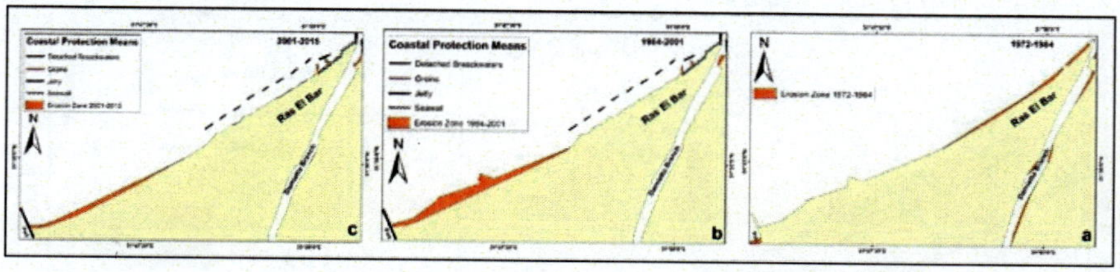

Fig. 5.10 Spatiotemporal changes in coastal erosion zones along Ras El Bar Beach

To protect the beach and the El Gameil Inlet from the coastal erosion hazard, five separate groins and a number of vertical barriers were built in the areas threatened by erosion, as shown in Fig. 5.11. At the entrance to the Suez Canal, two deep-sea jetties were used to protect the navigational passage of the canal from the risk of siltation inside the canal due to the movement of the current in the eastward direction.

5.2.1.3 Drain Inlets

Coastal erosion processes have affected the drain outlet and its navigational canals. In this study, three different drains were chosen to analyze the impact of coastal erosion on the outlet shoreline morphology, as shown in Fig. 5.12.

Abu Qir Drain

It is clear from analysis of the multitemporal satellite imagery of the AbuQir drain that the mouth of the drain was exposed to the risk of coastal erosion and lost an area of approximately 0.02 km^2 during the period between 2005 and 2009. The eroded area increased to 0.05 km^2 during the period from 2009 to 2017, which indicated an increase in the rates of coastal erosion on the downstream current, which needs urgent protection, as shown in Fig. 5.12.

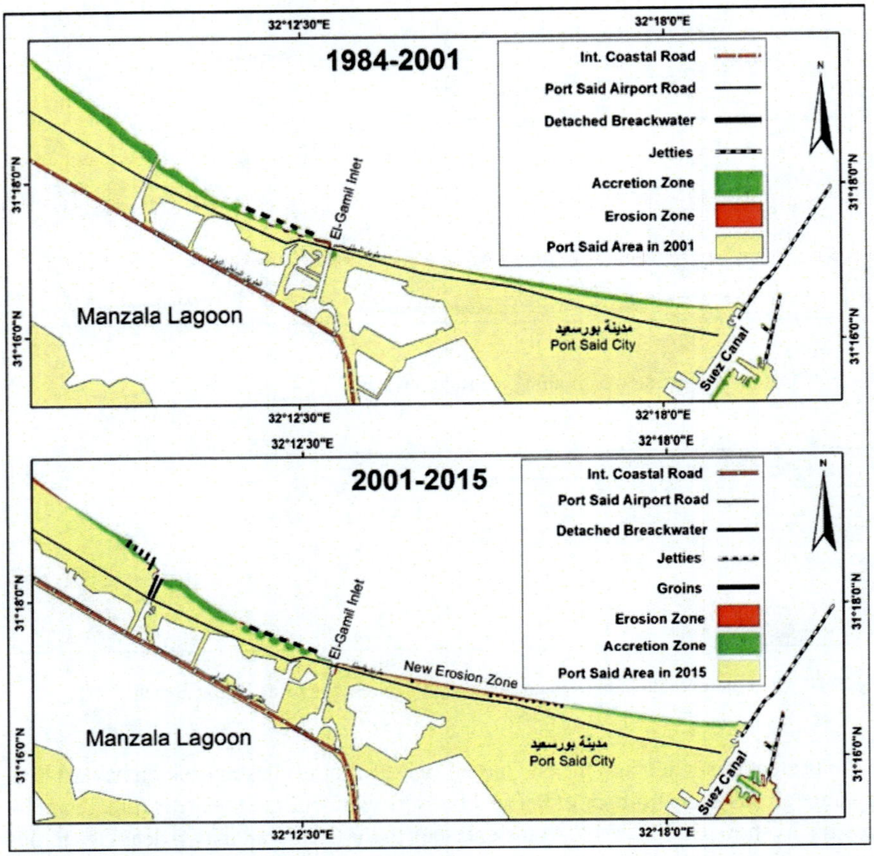

Fig. 5.11 Spatiotemporal changes in coastal erosion/accretion zones along Port Said Beach

Gharbia Drain

The mouth of the Gharbia Drain is one of the most exposed to coastal erosion in the study area, as it lost an area of approximately 0.46 km^2 during the period from 2005 to 2012 as a result of the coastal protection works along the western side of the promontory (vertical groins) that increased the rate of coastline retreat of the promontory. During the period from 2012 to 2017, the drain banks lost area and decreased to 0.21 km^2 due to the protection works along the promontory, as shown in Fig. 5.12.

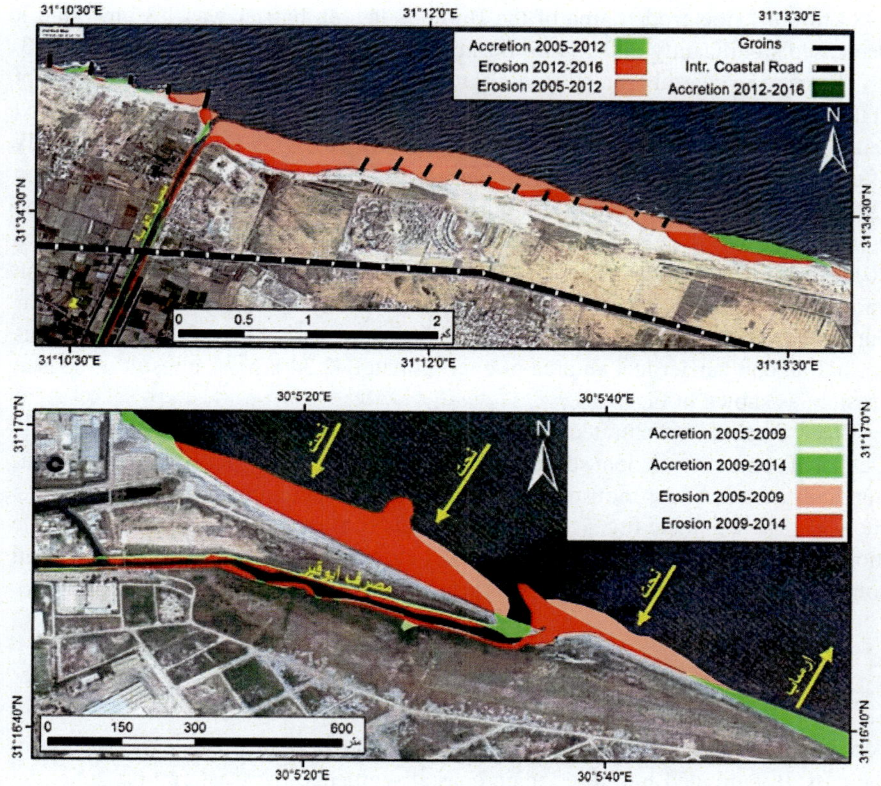

Fig. 5.12 Dynamic assessment of Coastline along Drains Inlets. **a** Gharbia Drain, **b** AbuQir Drain

Gamasa Drain

The area that was eroded from the downstream shoreline was 0.07 km^2 during the period between 2005 and 2009. The eroded area reached 0.06 km^2 during the consequent period 2007–2017 because of the construction of two groins in the mouth of the drain to protect it from retreat, but a new erosion zone along the eastern margin emerged and needs urgent protection, as shown in Fig. 5.12.

5.2.1.4 Coastal Lagoon Barriers and Black Sand Zones

Coastal erosion has significantly affected coastal lagoon barriers, and the comparison of multitemporal satellite remote sensing images and historical topographic maps indicated that the Burullus lagoon barrier lost an area of 1.83 km^2 during the period from 1945 to 1972, and the area increased to 4.56 km^2 after the construction of the High Dam during the period from 1972 to 1984. During the following period from

1984 to 2001, the eroded area of the Burullus lagoon barrier was 1.3 km^2, and it decreased significantly to 0.55 km^2 during the following period from 2001 to 2015.

There are no coastal protection works to protect the Burullus lagoon barrier except at the entrance of the Burullus inlet to avoid the movement of siltation and suspended sediments moving in the eastward direction, as shown in Fig. 5.13. Erosion negatively affected this barrier, and the barrier lost an area of 1.02 km^2 during the period between 1945 and 1972. It increased to 2 km^2 during the period from 1972 to 1984 and reached 2.16 km^2 during the period from 1984 to 2001, while during the period from 2001 to 2015, the eroded land was approximately 2.33 km^2. The Port Fouad coastal lagoon barrier was exposed to severe coastal erosion hazards, as it lost an area of 1.84 km^2 during the period from 1984 to 2001, while during the period from 2001 to 2015, this coastal lagoon barrier lost an area of approximately 2.34 km^2 as a result of coastal erosion, as shown in Fig. 5.13.

Black Sand is concentrated in large quantities on the northern coast of the Nile Delta and contains sufficient strategic and economic minerals for economic exploitation and use in nuclear industries as well as some other engineering industries. As shown in Table 5.4 and Fig. 5.13, these minerals contain many atomic elements, the most important of which are uranium, thorium, zirconium, hafnium, titanium, and some very rare earth elements, such as uranium oxide and thorium (Dabbour 1994).

5.2.1.5 Coastal Settlements

Alexandria city beaches extend from Dekheila port in the west to Abu Qir port in the east. It is divided into three main coastal segments: the western port zone, the eastern port, and the Alexandria Corniche. Analysis of coastal erosion zonation along Alexandria beach indicated that the eroded area between 1984 and 2002 was 1.09 km^2, while during the period from 2002 to 2015, the erosion area was estimated to be 1.12 km^2 from the coast, as shown in Fig. 5.14.

El-Sharnouby and Soliman (2011) divided the Alexandria Shoreline Protection System along the Alexandria shoreline into ten adjacent zones on different beaches:

- Zone (1) Shatby area: This area is a vertical seawall ranging between 3 and 5 m followed by a narrow sandy beach and was built to protect Corniche Road from the hazard of storms and severe coastal erosion, and the impact of coastal erosion on the front of the wall increased after the decline and disappearance of the sandy beach.
- Zone (2) Shatby to Cleopatra: This zone is 3500 m in length and is protected by a very hard structure (10 tons of concrete blocks) with a height of 5 m and a breadth of 20–30 m, and it is clear that many blocks break as a result of the continuous impact of storms (Fig. 5.15).
- Zone (3) extends along Cleopatra Beach and is protected by narrow Detached Breakwater with low peaks; these barriers have completely failed as the waves crossed the barrier, leaving only a small remnant.

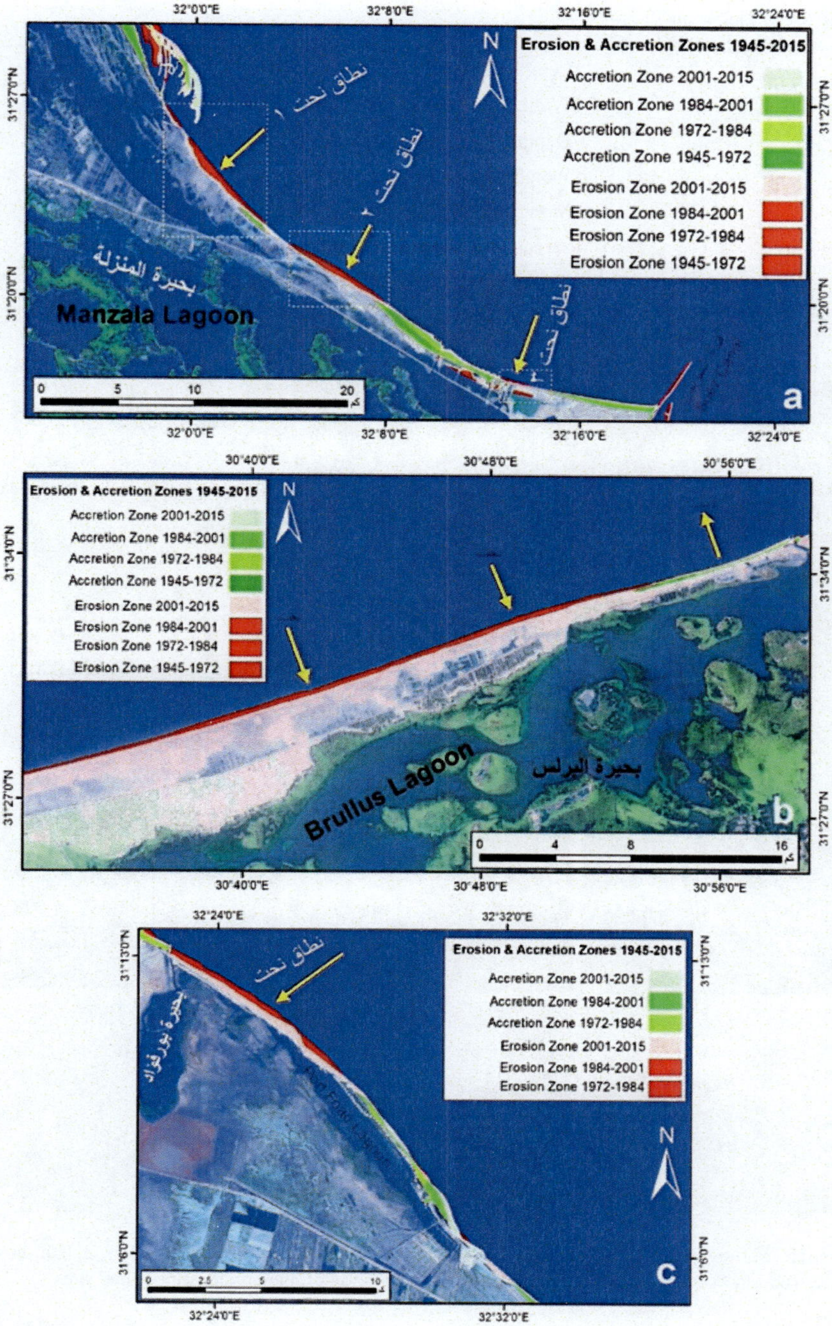

Fig. 5.13 Coastal erosion/accretion along coastal lagoon barriers. **a** Manzala, **b** Brullus, **c** Portfouad

Table 5.4 Black sand deposits along the NDC

	Rosetta	Baltim	Damietta
Length/km	70	15	30
Width/km	800	800	1000
Depth/km	9	15	10
Weight/Ton	839×10^6	306×10^{6}	510×10^{6}
Monazite/Ton	167.96	61.2	102
Zircon/Ton	1.847.560	673.2	1.122.000
Rutile/Ton	1.007.760	367.2	612

Source AfterDabbour (1994)

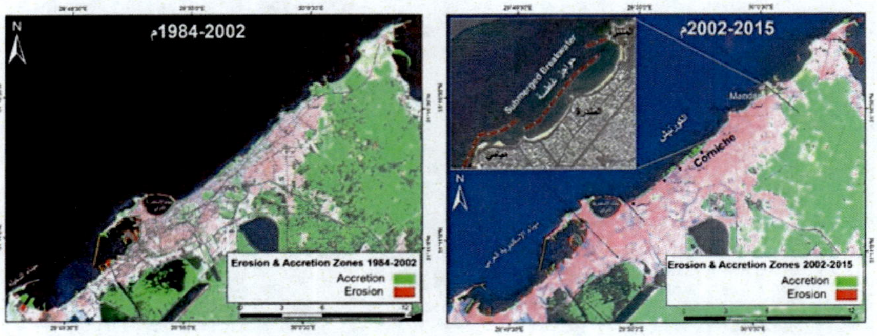

Fig. 5.14 Coastal erosion/accretion along the Alexandria Coast

Fig. 5.15 Photographs show the impacts of coastal erosion along Alexandria beaches. **a**, **b**, **c** show the destruction of buildings, and **e**, **f**, **g** show the destruction of ancient archaeological sites

- Zone (4) extends along Mustafa Kamel Beach with a length of 1500 m and is protected by a mixture of low- and high-top hills with concrete blocks.
- Zone (5) Historic Stanley Beach, which is protected by sand nourishment.
- Zone (6) Gleem beach, which is protected by grooves and vertical barriers, did not perform efficiently in the face of the 2010 storms, as they were crossed by waves.
- Zone (7) San Stefano Beach is surrounded by separate very hard barriers.
- Zone (8), which extends from San Stefano-El Mahrousa, is protected by concrete blocks.
- The area of (9) Sidi-Bishr is approximately 3000 m long, and there is no artificial protection thus far due to the presence of submerged natural coral reefs (Fig. 5.15).
- Zone (10) The zone of the beach extending from Miami—the park with a length of 3500 m and this area suffered from severe Coastal Erosion due to storms in 2003, Submerged Breakwater submersible barriers were built during the period from 2006 to 2009, and the barrier consists of three parallel parts with a length of 3000 m (150–350 m in deep water at a depth of 2.5–8.5 m) and the beach width ranges between 36 and 46 m, The barrier was exposed to the impact of storm surges in 2010 (El-Sharnouby and Soliman 2011).

Al-Burj village along the Brullus headland is one of the most important coastal settlements threatened by coastal erosion hazard and risk. Spatiotemporal analysis of the coastline indicated an obvious retreat range between 65 and 298 m in front of the village of Al-Burj during the period between 1945 and 2015. To protect the village of Al-Burj, the coastal protection agency established a 1.2 km-long seawall to the east of the Brullus inlet, consisting of Basaltic Riprap and Concrete Revetments (Fig. 5.16).

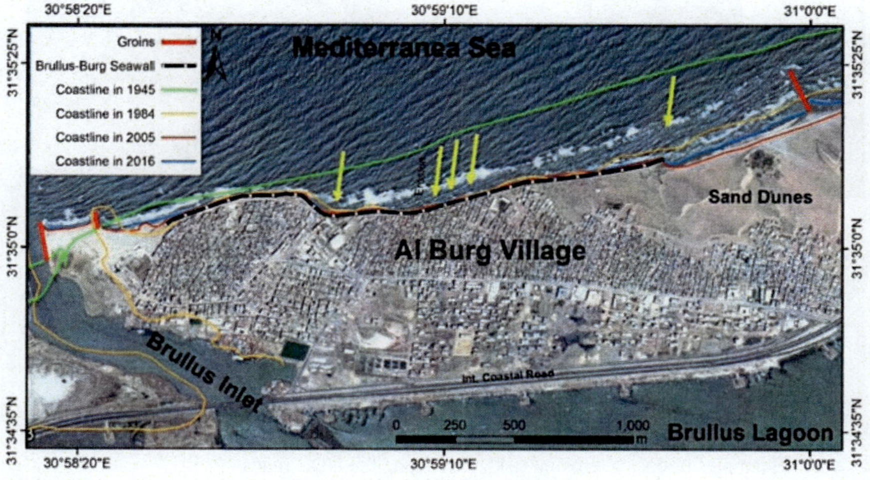

Fig. 5.16 Coastline dynamic along Al Burg–Brullus headland

5.2.1.6 Coastal Roads

Coastal erosion processes threaten the coastal roads along the NDC. The analysis of satellite imagery and field studies indicated that four different sites are vulnerable to severe coastal erosion, which are the Alexandria-Dekheila road at the El-Max area on the western side of Alexandria, approximately 191.9 m from the road to the shoreline in 1945, while the distance reached 15 m from the road to the beach line in 2016. The international coastal road east of Al-Burj village. The distance from the road to the shoreline was 435 m in 1984, while it reached 357 m in 2016. The Coastal Erosion affects the coastal road between Damietta and Port Said, as the distance from the road to the beach line reached 1489.9 m in 2004, while the distance from the road to the shoreline reached 16 m in 2016. The Coastal Erosion affects the airport road in Port Said, as the distance from the road to the beach line reached 99.8 m in 2001, while the distance from the road to the beach line reached 40.4 m in 2016 after the construction of a coastal protection wall, as shown in Fig. 5.17.

5.2.2 Siltation Risk Assessment

Siltation of suspended sediments that resulted from coastal erosion and transported downdrift may cause large problems in navigational channels and streams. In this study, a detailed study of sediment accumulation in the streams of the Nile River estuaries, lagoon inlets, ports, and navigational canals was conducted.

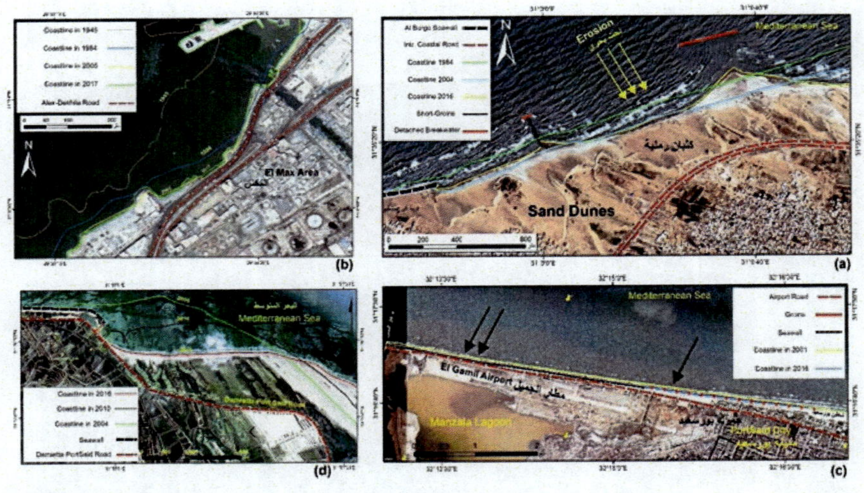

Fig. 5.17 Examples of coastal erosion risk zones along the NDC roads

- Hazard of coastal sedimentation in the waterways of the mouths of the Nile River, it is clear from the interpretation of the satellite imagery of the Rosetta promontory channel from 1984 to the present that there is a sandy tongue formed in the promontory course with an area of 0.39 km^2 in 1984, its area increased significantly to 0.78 km^2 in 1990, then its area decreased to 0.089 km^2 and its hazard increased in the year in 2002, he worked on a semi closure of the promontory, and the problem continued until 2005, when its area reached 0.098 km^2, while in 2017 it became clear that the sand tongue moved toward the south inside the stream, and its area increased again to 0.14 km^2, and the sand tongues obstruct the navigational movement in the important navigational streams as shown in Fig. 5.19.
- Similar to the mouth of the Rosetta branch, it is clear that the hazard of sand spit expanded to the navigation channel at the mouth of the Damietta branch, as the area of the sand spit in 1984 reached 0.06 km^2, and its area decreased in 2010 to 0.051 km, as evidenced by the analysis of satellite images in Figs. 5.18 and 5.19.

Sedimentation processes cause problems in the coastal Lagoons Inlets, and the Egyptian government is protecting both sides of the Suez Canal to stabilize the navigational streams from siltation in the navigational paths.

- ***Edku Lagoon Inlet***: Two jetties were constructed on both sides of the Idku Inlet in 1982 to protect the navigational canal from continuous sedimentation resulting from the movement of sediment toward the east. It is 240 m and 175 m long on the west and east sides, respectively (Frihy et al. 1991). The results from the analysis of satellite images during the period from 2002 to 2017 indicated that a sedimentary zone formed behind the western jetty with an area of 0.108 km^2, where the jetty succeeded in impounding sediment away from the navigational canal of the Inlet, as shown in Fig. 5.20.
- ***Burullus Lagoon Inlet***: The western jetty was built on Burullus Inlet between 1971 and 1972, and most of the sediment turned around it due to its small length. The second jetty was built in the direction of the east in 1983 at a distance of 250 m. However, sediment began to accumulate between the two jetties, which hindered navigation in the inlet. The third jetty was built to discourage navigational paths and establish a fishing port. The analysis of satellite observations during the period between 2005 and 2017 shows that there is a sedimentary zone of 0.217 km^2 behind the western jetty of the inlet, and these jetties succeeded in impounding sediment and protecting the inlet from sedimentation processes in the eastward direction, as shown in Fig. 5.20.
- **Manzala Lagoon Inlet**: The analysis of the up-to-date satellite images indicates that the El Gameil Inlet is exposed to Coastal Erosion and sedimentation hazard and risk. The inlet was protected by five vertical barrier grooves, which were built during the period from 2002 to 2004 to protect the inlet from coastal erosion. There are also two jetties to protect the navigational course from the risk of sedimentation. From the analysis, the erosion zone was 0.05 km^2, and a sedimentation zone of 0.26 km^2 was behind the vertical barriers on the West inlet, as shown in Fig. 5.20.

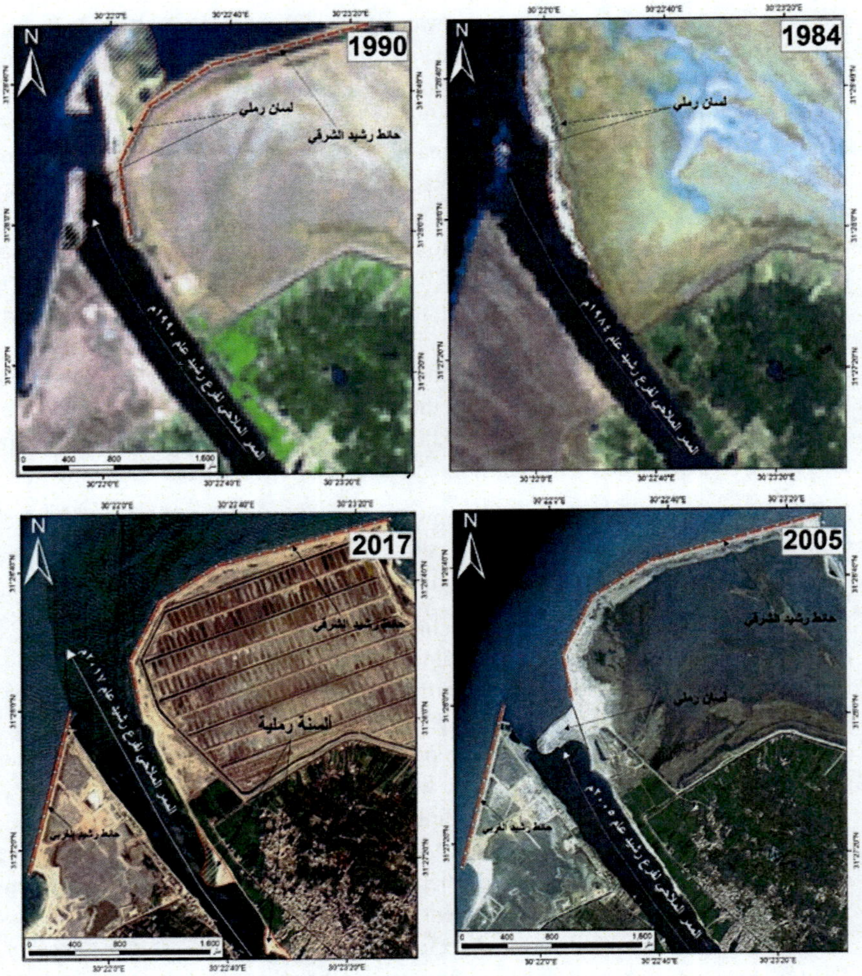

Fig. 5.18 Examples of coastal siltation problems along NDC

5.2.2.1 Hazard of Sedimentation in the Port

Sedimentation processes threaten the streams of navigational canals and important ports in the study area, an analytical study of the hazard of sedimentation in the navigational canal of the Suez Canal and the new port of Damietta.

- Suez Canal Inlet

The longest jetties in the study area are on the western side of the Suez Canal with a length of 7.7 km seaward, and it was constructed in 1864 to protect the canal from sediments coming with the permanent current eastward. This jetty worked to accumulate sediments behind it, forming a sedimentary sand beach in front of the

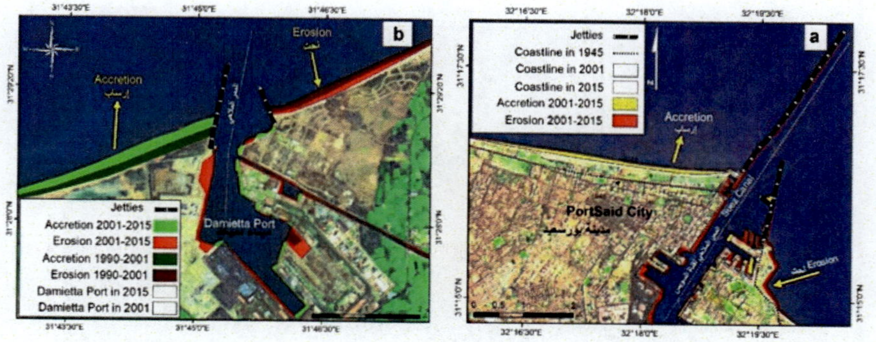

Fig. 5.19 Examples of sedimention and siltation problems of the NDC navigational channels and ports

city of Port Said. The analysis of erosion and sedimentation zones along both sides of the canal showed that a sedimentary zone of 0.47 km^2 was behind the western jetty during the period from 2001 to 2015, while the erosion zone on the banks of the entrance to the canal had an area of 0.71 km^2 during the period from 2001 to 2015, as shown in Fig. 5.20.

- New Damietta Port

Two jetties were built on both sides of Damietta Port in 1985 with lengths of 1300 m for the western jetty and 600 m for the eastern jetty to prevent and reduce sedimentation processes in the port's navigational channel. The analysis of the erosion and sedimentation zones along both sides of the port indicated that the sedimentation zone occurred behind the western jetty by approximately 0.65 km^2 during the 1990–2001 period and increased to 0.66 km^2 during the 2001–2015 period. Along the eastern side of the eastern jetty, a new erosion zone (0.57 km^2) formed during the 1990–2001 period, which decreased to 0.55 km^2 during the 2001–2015 period, as shown in Fig. 5.20.

5.3 Evaluation of the Effectiveness of Coastal Protection Structures

The United Nations report in cooperation with the Egyptian Coastal Protection Agency and the Coastal Research Institute in Alexandria indicates that the protection of NDC during the period 2009–2014 due to coastal erosion and climate change impacts has spent a budget of $16,838,060. These funds have been used in the maintenance and renovation of concrete protection seawalls and the construction of short groins and detached breakwaters (UNDP 2009).

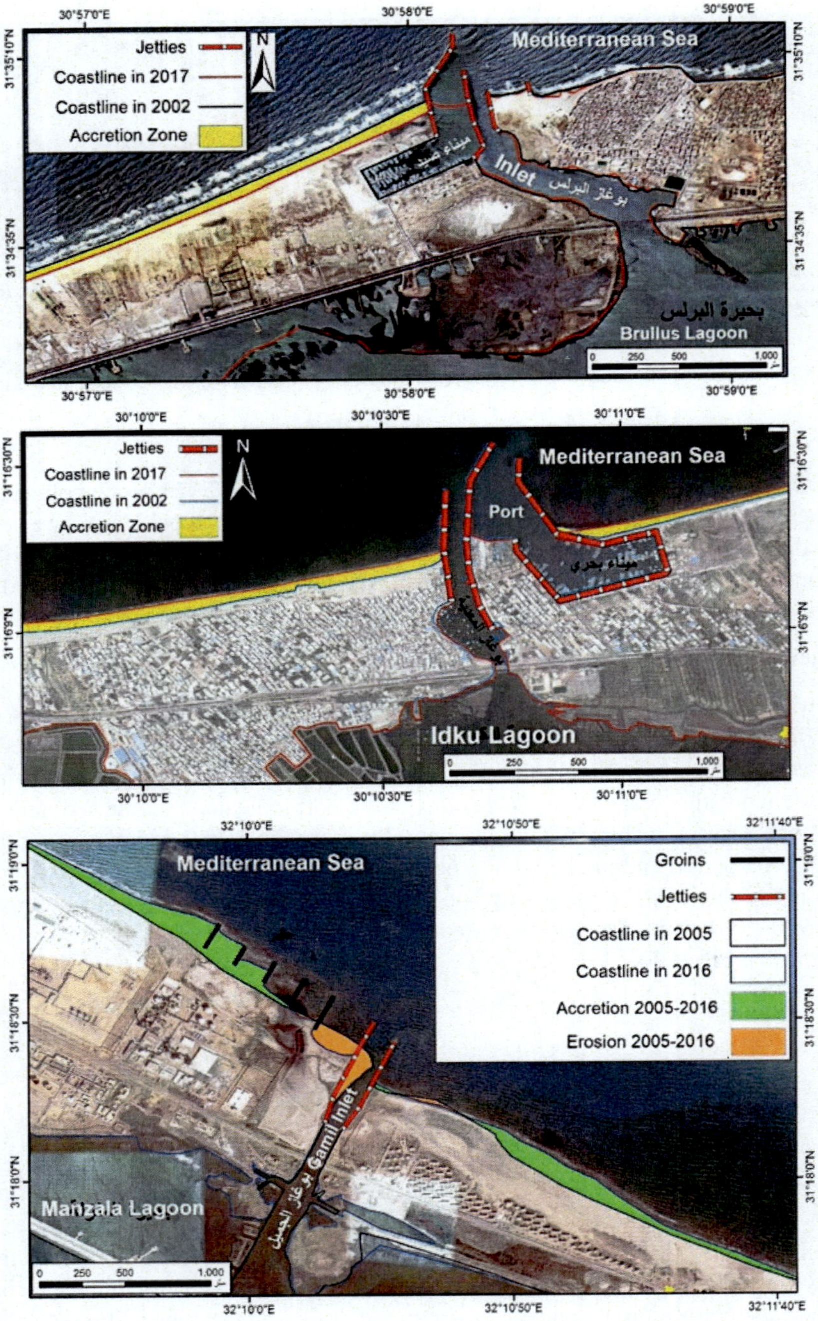

Fig. 5.20 Examples of sedimentation and siltation problems along coastal lagoon inlets

Table 5.5 Spatial changes in Rosetta promontory annual erosion rates

Period	Area (sq.m)	Area (sq.km)	Annual erosion rate (m2/yr.)	Annual erosion rate (km2/yr.)
2001–2015	3,013,524.31	3.01	215,251.74	0.22
1984–2001	6,726,978.03	6.73	395,704.59	0.4
1972–1984	8,642,042.33	8.64	720,170.19	0.72
1945–1972	4,388,374.9	4.39	162,532.4	0.16

5.3.1 Rosetta Promontory

The assessment of coastal protection structures along the Rosetta promontory shows that coastal protection seawalls have succeeded in reducing annual coastline retreat and land degradation due to coastal erosion. According to Ghoneim et al. (2015), the Rosetta promontory retreated 2.3 km landward, and it would have lost an area of 7.2 km^2 without the construction of coastal protection walls in the downstream direction.

From field observations, it is clear that some concrete blocks of these seawalls have sunk and disintegrated due to salt weathering, which prompted the Egyptian Shore Protection Agency to establish a factory near the mouth of Rosetta to manufacture triple- and quadruple-block tetrapod has to feed the seawalls. An analysis of the hazards of coastal erosion along both sides of the promontory after the construction of coastal protection means shows the following:

1. New erosion zones (3.7 km long) began to appear on the margins of protected zones on the eastern side (end of the short vertical barrier groins) and (1.8 km on the western side).
2. During the post protection period 2001–2015, due to coastal erosion, the promontory lost an area of 3.01 km^2 with an annual loss rate of 0.22 km^2/year, as shown in Table 5.5 and Fig. 5.21.

5.3.2 Baltim-Burullus Zone

Although coastal protection engineering works have succeeded along the Baltim resort area (detached breakwaters, short groins) and in front of the village of Al-Burj (seawall), which worked to reverse the system of erosion into sedimentation, new areas of erosion have appeared on the margins of the protected areas, and the rate of erosion is increasing annually and needs protection as well. It is clear from Figs. 5.22 and 5.23.

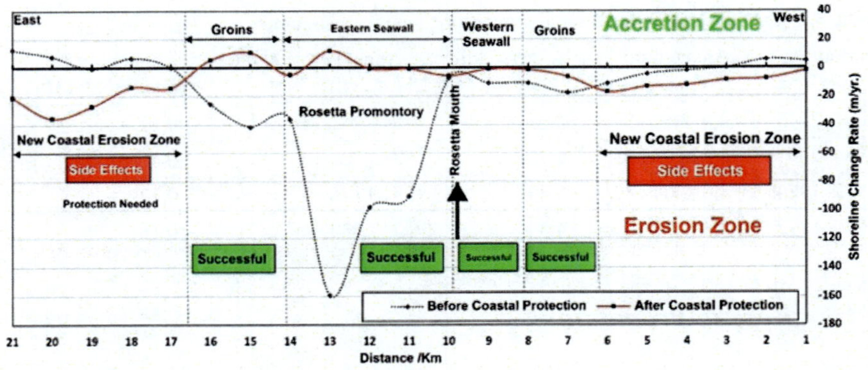

Fig. 5.21 Show coastal protection evaluation along the Rosetta promontory

Fig. 5.22 Examples of the effectiveness of coastal protection works along Rosetta seawalls

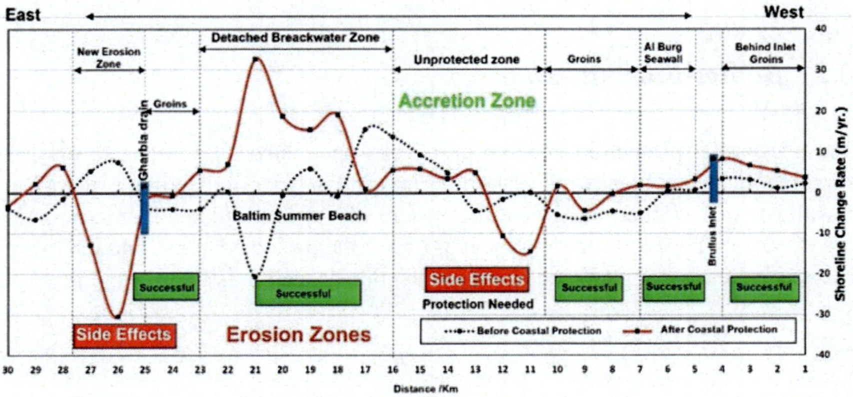

Fig. 5.23 Show coastal protection evaluation along the Brullus headland

5.3.3 Ras El Bar-Damietta Promontory

The coastal protection system along the Ras El Bar area (groins, detached breakwaters, jetties) and in front of the mouth of Damietta (eastern Damietta seawall) worked to protect the coastal zone of the promontory and the resort from severe coastal erosion risk. On the other hand, there are side effects produced as a result of coastal protection works, but new zones of erosion have appeared on the margins of the protected zones on the eastern side of the protection seawall and on the western side of the resort, and the rate of erosion increases annually and needs protection, as shown in Figs. 5.24 and 5.25.

Fig. 5.24 Failure of coastal protection works along the Brullus and Damietta seawalls

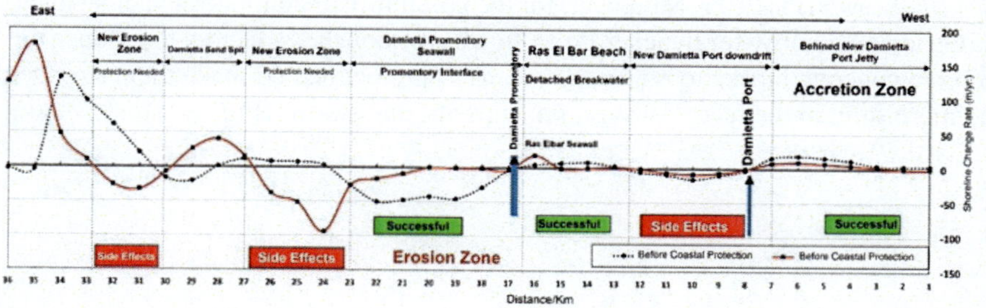

Fig. 5.25 Show coastal protection evaluation along the Damietta Promontory

Natural factors affect coastal protection structure efficiency, the most important of which are as follows.

5.3.4 Underwater Sunken of Coastal Protection Structure

It is clear from the field studies in the study area that there are many protection means drowned under the sea, the most important of which are the coastal seawall in the Burullus area, the detached breakwaters in the Ras al-Bar area, the short groins in the Damietta area, and some separate barriers in front of the Alexandria coast that were sunk by storm surges, as shown in the photographs (Fig. 5.24).

Many coastal protection means have been recorded that have been buried due to the accumulation of sediments around them. The analysis of these means indicated that many of these means have been constructed in the wrong locations that approach the sedimentation zones, which worked to bury them in modern sediments, the most important of which are the separate barriers in front of the Baltim resort, which were buried under the modern sediments due to their location in the sedimentation zone. For landfill in sediments that are more than four meters high. Disintegration and breaking of coastal protection means due to weathering and marine factors. Many protection means are subject to mass disintegration and breakage because of their inability to withstand the action of waves, especially during the winter season. Many cases have been recorded in the study area. For example, in the Shehabia area in Baltim, the concrete blocks that make up the separate barriers were subjected to decomposition into their basic components represented in gravel of varying size, as shown in Fig. 5.23. For the Rosetta Wall, the concrete blocks were subjected to cleavage and bulk disintegration due to mechanical and chemical weathering. The effect of erosion and sea traction appears because of the impact of strong storms on the cubic-shaped concrete blocks in the Al-Max-Dekheila area. Maintenance of means of protection, replacement, and renewal using concrete blocks of varying size, time of installment, and shape, where the new means of protection consist of basaltic fractions and concrete blocks on a regular basis, according to the extent of the hardness of the concrete blocks in front of the action of waves and storms. Refeeding with stronger and more solid concrete blocks at the Chinese Seawall (Rosetta Seawall) in 2011. Moreover, for the possibility of continuous maintenance, the General Authority for Beaches Protection has established a factory to produce the required concrete blocks to renew the means of protection and make them stronger and more solid in the face of waves, sea currents, and storms (Fig. 5.24).

5.3.5 Shoreline Positional Prediction

The future shoreline position can be spatially calculated using the following equations:

Shoreline position $=$ slope (MEPR) $*$ date interval (T) $+$ intercept
Predicated Shoreline position (P) $=$ MEPR $*$ T $+$ BEPR
Slope (MEPR) $=$ (P1 $-$ P2)/(T2 $-$ T1)
Intercept (BEPR) $=$ P1 $-$ (MEPR $*$ T1) $=$ P2 $-$ (MEPR $*$ T2)(Nandi et al., 2016, p. 66)

where:
MEPR = EPR shoreline change rate.
T = The period of time between the current shoreline and the desired shoreline in the future.
BEPR = Intersection of the values of the prediction axis with the Y axis and called Intercept.
P1 = Older shoreline.
P2 = Newer shoreline.
T1 = Older time period.
T2 = Most Recent Time Period.

5.4 Coastal Vulnerability Index (CVI)

Coastal hazards are increasing in many regions along the world because of the rapid change in natural, anthropogenic, and geological variables that cause and increase dynamic coastal processes (Mujabar and Chandrasekar 2013). The coastal vulnerability index (CVI) is used to map relative risk along the study area based on GIS techniques; the index depends on giving ranks to natural, marine, and human variables that affect the degrees of risk in the study area. The data used in the CVI index vary according to the degree of importance according to the spatial scale used. Hazards occur because of the interaction between many human and natural coastal processes.

1. Physical variables

Natural variables include geology, geomorphology, the action of currents and waves, spatial variations in sea level rise, landslides, elevations, and slopes.

2. Human variables

Human activities affect hydrological cycles along beaches, including coastal protection works and population density. To apply the CVI along the NDC, the study area was divided into an equal network of grids covering the coastline with dimensions (1 × 1 km) with a total grid of 543 cells. Eleven variables affecting the degrees of risk in the study area were identified and collected in a geodatabase (geological sediments, geomorphological units, shoreline change rate, relative SLR, land subsidence, coastal protection, average wave height, daily tide zone, topography, slope, and population density). Each cell has a code registered within the spatial database table. The process of dividing the coast into parts or networks is one of the fundamentals of coastal hazard analysis and modeling (Kairu and Nyandwi 2000). The study area was classified into five geomorphological zones: the Hammam-Alexandria zone, the Abu Qir Bay-Rosetta promontory, Burullus headland zone, the Damietta promontory zone, and the Manzala-Port Said zone. Table 5.6 shows the natural characteristics of the geomorphological zones, and Fig. 5.26 shows the system of cells covering the coastline.

5.4.1 Calculation of CVI

The coastal vulnerability index (CVI) is one of the world's most well-known direct methods for assessing the coastal risk of sea level rise and coastal erosion; it was developed by Gornitz et al. (1994) to assess the coastal risk of sea level rise and coastal erosion. In the CVI, there is a set of variables that represent the forces acting along the coast, which they are mathematically multiplied to obtain an index that can then be classified according to the value of the index. Thieler and Hammar-Klose (1999) modified the CVI index by reducing the number of variables to become more effective, practically applicable, and more influential in their study.

The US Geological Survey has adopted that the internationally approved CVI index becomes at least six variables, namely, geomorphology—change in shoreline—coastal slope—relative sea level height—average wave height—and average tidal range, and Eq. (5.1) becomes as follows:

$$\mathbf{CVI} = \sqrt[2]{\frac{\mathbf{a} * \mathbf{b} * \mathbf{c} * \mathbf{d} * \mathbf{e} * \mathbf{f} * \mathbf{g}}{6}} \quad (5.1)$$

a = geomorphology; b = coastal slope (%); c = relative sea level rise (mm/yr.); d = Shoreline erosion/accumulation rate (m/yr.); e = mean tide range (m); and f = mean wave height (m).

In this study, the coastal hazard index in the study area was modified, including eleven variables that were entered into the database and encoded in CVI-based ESRI's ArcGIS. The hazard index in the study area was calculated according to the following Eq. (5.2). Figures 5.26 and 5.27 shows the calculation method using ArcGIS:

Table 5.6 Physical criteria effects on the coastal vulnerability index along the NDC

Variables	Alexandria	AbuQuir Bay	Brullus Headland	Damietta Promontory	Port Said
Geology	Recent Quaternary Deposits (Sand and Gravel), Nile Silt, Rocky Carbonate Beach Ridges Sabkhas, Lagoons, Stabilized Dunes	Recent Quaternary Deposits (Sand), Sand Dunes, Sabkhas, Lagoons, Stabilized Dunes, Nile Silt	Recent Quaternary Deposits (Sand), Sand Dunes, Sabkhas, Lagoons, Nile Silt	Recent Quaternary Deposits (Sand), Sand Dunes, Sabkhas, Lagoons, Nile Silt	Nile Silt, Stabilized Dunes, Sabkhas
Geomorphology	Carbonate Ridges, Nile silt	Nile silt, Beaches and Backshore, Sabkha	Beaches, Aquaculture, Lagoons, sabkhas, Sand Dunes	Beaches, Aquaculture, Fish Farms Lagoons, sabkhas, Stabilized Dunes	Beaches, Fish Farms Lagoons, sabkhas, Desert
Shoreline Configuration	SW–NE	NW–SW, SW–NE	SW–NE, NW–SW	SW–NE, W–E, NW–SE	NW–SE
Beach Composition	Very Coarse to coarse grains, Limestone Fractions, gravels	Very Fine to Fine Grains sand, Black sand, silt, Salt	Medium grains sand, Black sand, Mud, silt Salt, Sand dunes	Fine grains, Silt, Salt	Medium to coarse Grain Sand, Salt
Wave Height (m)	Max: 1.67 m Min: 0.1 m Avg:0.37 m	Max: 6.82 m Min: 0.11 m Avg:1.26 m	Max: 6.02 m Min: 0.11 m Avg:0.64 m	Max: 4.63 m Min: 0.12 m Avg:0.93 m	Max: 2.22 m Min: 0.12 m Avg:0.58 m
Tide Gauge (Not-significant)	Microtidal Environment				
Beaches Width (m)	50–200 m	10–100	50–4000	30–1000	30–2000
Drainages/ Estuaries	Drains	Rosetta Promontory, IdKu outlet	Brullus Outlet, Drains	Damietta Promontory, Drains	Suez Canal, Manzala Outlet

$$\mathbf{CVI} = \sqrt[2]{\frac{\mathbf{a} * \mathbf{b} * \mathbf{c} * \mathbf{d} * \mathbf{e} * \mathbf{f} * \mathbf{g} * \mathbf{h} * \mathbf{i} * \mathbf{j} * \mathbf{k}}{11}} \tag{5.2}$$

where:

a = Geology; b = Geomorphology c = Shoreline Change Rate (m/year); d = Topography (m); e = Coastal Slope (deg); f = Relative Sea Level Change (mm/year); g

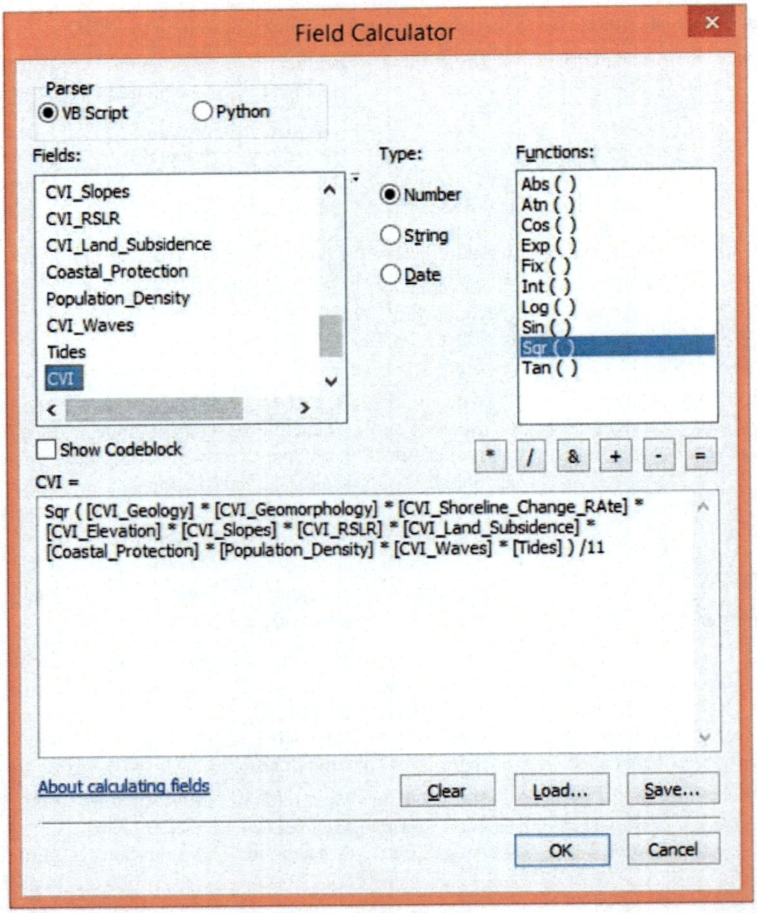

Fig. 5.26 Calculation process of CVI using the ArcGIS field calculator

= Land Subsidence (mm/year); h = Coastal Protection; i = Population Density (capita); j = Mean Wave Height (m); k = Mean Tide Range (m).

The analysis of the CVI results for the study area shows the following:

1. Very high-risk areas occupy 20.89 km out of a total of 415.35 km on which the index was implemented by 5.03% and appear along the Rosetta promontory and the Abu Qir Bay area at the Muhammad Ali Wall, as shown in Figs. 5.28, 5.29, 5.30, 5.31.
2. High-risk areas occupy 70.10 km, or 16.88% of the length of the study area and are distributed along the Damietta promontory and some coastal zones on the Burullus checkpoint, on both sides of the Rosetta promontory, the mouth of the Gharbia Drain, and the Manzala coastal lagoon barrier at Barr al-Dahra.

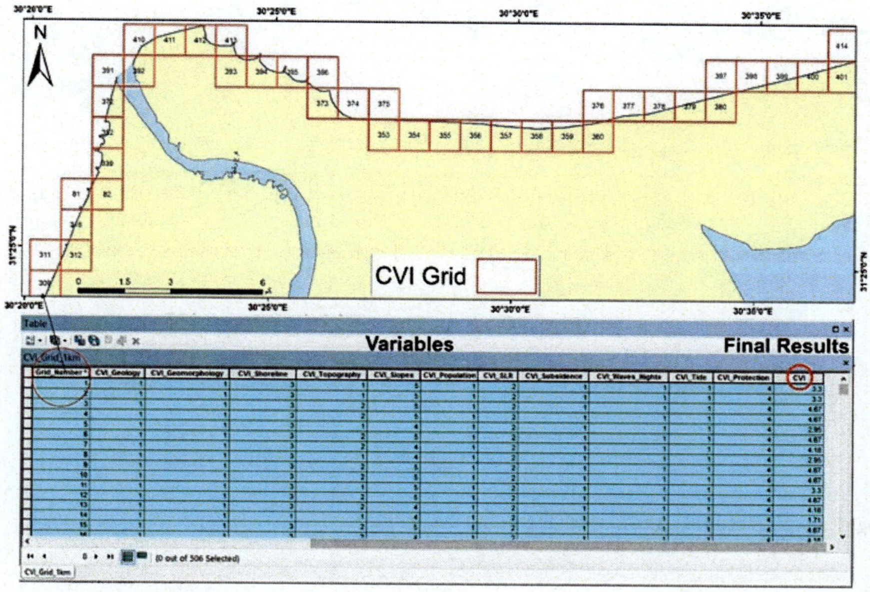

Fig. 5.27 GIS-based CVI grid system

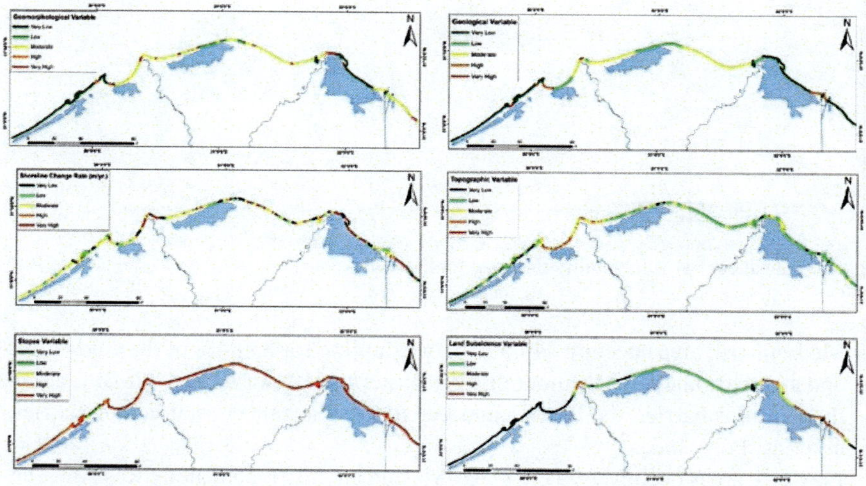

Fig. 5.28 Physical criteria risk zones according to CVI grids

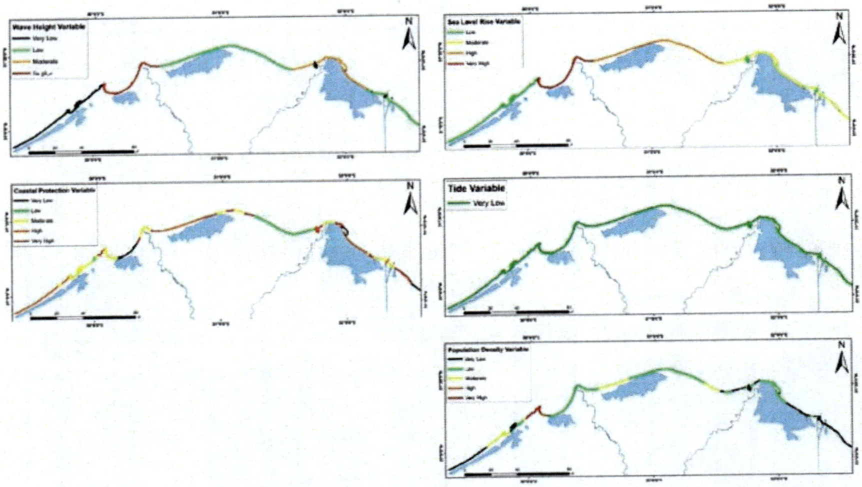

Fig. 5.29 Human criteria risk zones according to CVI grids

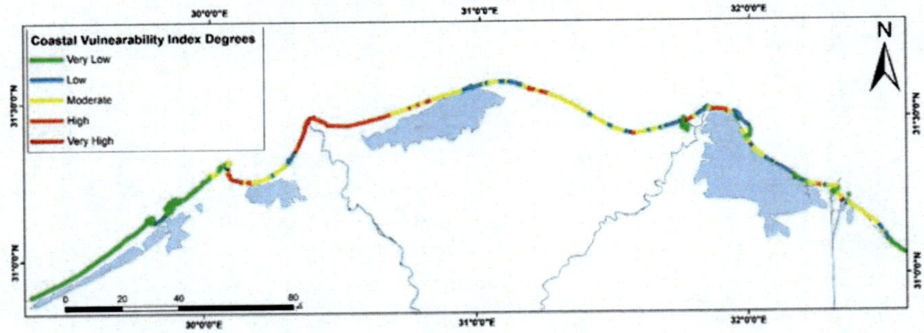

Fig. 5.30 Final coastal vulnerability risk map for the study area

3. Medium-risk areas occupy 140.55 km, accounting for 33.84% of the study area, and are distributed in Maamoura Bay in eastern Alexandria, Abu Qir Bay, along the Burullus barrier, the East Gamasa area, the Manzala coastal lagoon barrier, and East Port Said.
4. Low-risk areas occupy 1.99 km by 0.48% and are distributed along Ras Abu Qir, Burullus Beach, Gamasa Beach, Manzala coastal lagoon barrier, and some areas of Port Said Beach.
5. Very low areas (Stable Zone) occupy a distance of 181.82 km by 43.77% and are distributed along the Hammam-Alexandria range, Damietta sandy spit, and some zones of the Manzala coastal lagoon barrier, as shown in Figs. 5.27, 5.28, 5.29, 5.30 and Table 5.7.

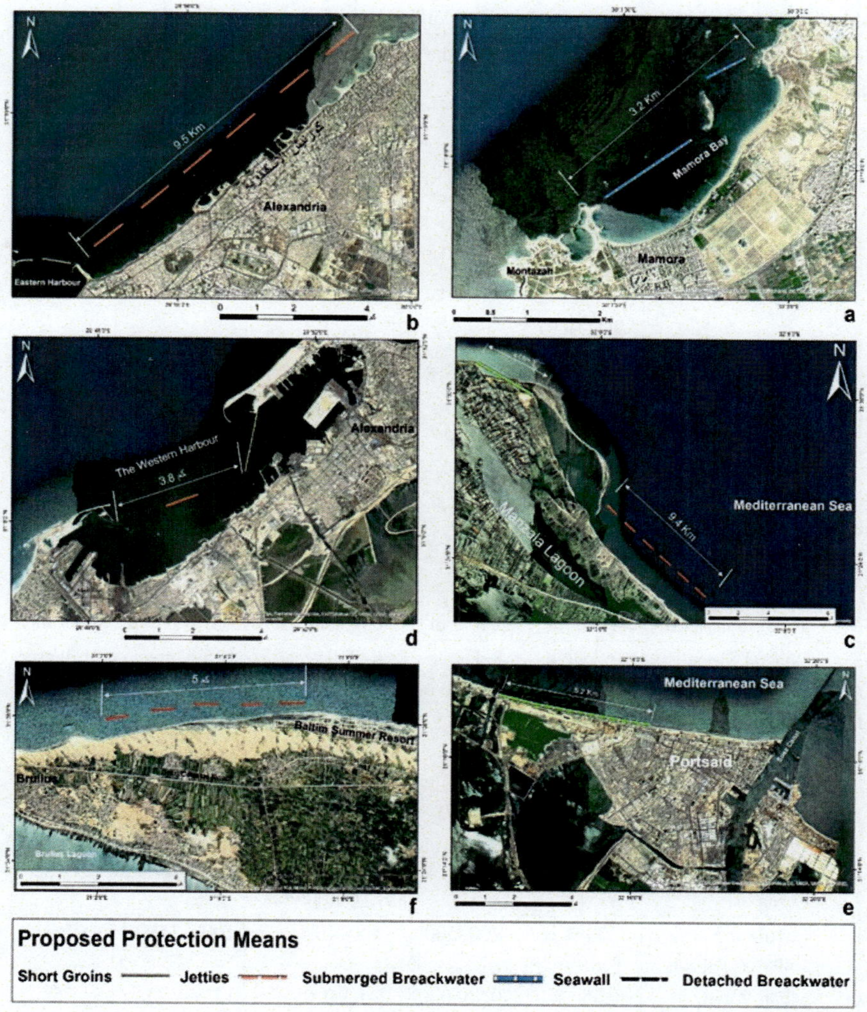

Fig. 5.31 Proposed coastal protection works along the high-risk coastal zone

5.5 Summary

This chapter addresses the study of coastal erosion hazards and risks along the NDC using satellite remote sensing and GIS techniques. Detailed studies for coastal erosion and sedimentation mapping, assessment, and management along the NDC. The study area was geomorphologically divided into five coastal zones, including the Alaxandria-hamam zone, the Rosetta promontory, Brullus headland, Damietta promontory, and Manzala lagoon to Port Said coastal zone. Moreover, the evaluation of the engineering coastal protection works that were built to protect the Nile delta

Table 5.7 Human criteria effects on the coastal vulnerability index along the NDC

Zones for vulnerability ranking of variables on the study area						
No.	CVI- Risk values	Very Low	Low	Moderate	High	Very High
	GIS-Geocode	1	2	3	4	5
1a	Geology	Stabilized Coastal dunes, Beach Ridges, Recent Quaternary deposits "Gravels"	Sand Dunes	Recent Coastal Deposits "Sand"	Nile Silt	Sabkhas, and Salt March
2b	Geomorphology	Carbonate Ridges, Stabilized Dunes	Sand Dunes	Beach sand, and Backshore	Nile silt	Sabkhas, and Salt March
3c	Shoreline Change Rate (m/year)	> +5.1	(+5) – (+2.1)	(-2) – (+2)	(−2.1) – (−5)	< −5.1
		Accretion		Stable	Erosion	
4d	Topography (m)	>4 m	2 to 3 m	0 to 1 m	−1 to −2 m	< −3 m
5e	Coastal Slope (deg)	>9 degree	8 degree	6–7 degree	1–5 degree	<1 degree
6f	Relative Sea Level Change (mm/year)	<1.8	1.8–2.5	2.5–2.95	2.95–3.16	>3.16
7g	Land Subsidence (mm/year)	<1 mm/y	1 mm/y	2 mm/y	2–3 mm/y	>4 mm/y
8h	Coastal Protection	Natural Protection		Engineering Construction	Unprotected	
9i	Population Density (capita)	<200	200–500	500–1000	1000–1500	>1500
10j	Mean Wave Height (m)	<0.55	0.55–0.85	0.85–1.05	1.05–1.25	>1.25
11k	Mean Tide Range (m)	>6	4.1–6	2–4	1–1.9	<1

promontories from severe coastal erosion, which dramatically increased after the construction of the High Dam in 1964, beaches of resorts such as Baltim and Ras El Bar, Port Said, and Alexandria beaches. In addition, coastal erosion hazards were assessed along coastal settlements, drains, canals, ports, and coastal roads. In addition, coastal protection effectiveness was evaluated along high-risk coastal erosion sectors. The study concluded that there are new zones of coastal erosion risk zones that began to appear on the eastern and western margins of the protection zones represented in the coastal seawalls and short groins barriers, which follow the direction of the movement of the Littoral subcells transport current. The coastal hazard index (CVI) was established for the study area based on eleven variables (geological sediments, geomorphological units, coastline changes, relative sea level rise, subsidence, coastal protection, average wave height, daily tidal range, topography, and slope, and population density).

References

Chen Z, Warne AG, Stanley DJ (1992) Late quaternary evolution of the northwest Nile delta between Rosetta and Alexandria, Egypt. J Coast Res 8:527–561

Dabbour GA (1994) The Egyptian placer deposits—a potential source for nuclear raw materials. In: Second Arab Conference on the Peaceful Uses of Atomic Energy, Cairo 5–9 November 1994, pp 191–204

Darwish K, Smith S, Torab M, Monsef H, Hussein O (2017) Geomorphological changes along the Nile Delta coastline between 1945 and 2015 detected using satellite remote sensing and GIS. J Coastal Res 33(4):786–794

Di K, Wang J, Ma R, Li R (2003) Automatic shoreline extraction from high-resolution Ikonos satellite imagery. In: ASPRS Annual Conference Proceedings. May 2003, Anchorage, Alaska

El-Sharnouby B, Soliman A (2011) Behavior of shore protection structures at Alexandria, Egypt, during the storm of December 2010. In: The International Maritime Transport and Logistics Conference "A Vision For Future Integration" 18–20 December 2011

Fanos AM, Frihy OE, Khafagy AA, Komar PD (1991) Processes of shoreline change along the Nile Delta coast of Egypt. Coast Sediments Conf 2:1547–1557

Fanos AM, Khafagy AA, Dean RG (1995) Protective works on the Nile Delta Coast. J Coastal Res 11:516–528

Ferraras SF, Pararas-Carayannis G (1988) Natural and man-made coastal hazards. In: Proceedings of the 3rd international conference on natural and man-made coastal hazards held in August 15–20, 1988. Ensenada, Baja California, pp 37–42

Frihy OE, Debes A, El Sayed W (2003) Processes reshaping the Nile delta promontories of Egypt: pre- and post protection. Geomorphology 53:263–279

Frihy OE, El Banna M, El Kolfat A (2004) Environmental impacts of Baltim and Ras El Bar shore-parallel breakwater systems on the Nile delta littoral zone, Egypt. Environ Geol 45(3):381–390

Frihy OE, Fans MA, Khafagy AA, Komar PD (1991) Nearshore sediment transport patterns along the Nile Delta, Egypt. Coastal Eng 15:409–429

Frihy OE, Lawrence D (2004) Evolution of the modern Nile delta promontories: development of accretional features during coastline retreat. Environ Geol 46:914–931

Ghoneim E, Mashaly J, Gamble D, Halls J, Abu-Bakr M (2015) Nile Delta exhibited a spatial reversal in the rates of shoreline retreat on the Rosetta promontory comparing pre- and post-beach protection. Geomorphology, 228:1–14. https://doi.org/10.1016/j.geomorph.2014.08.021

Gornitz VM, Daniels RC, White TW, Birdwell KR (1994) The development of a coastal vulnerability assessment database, vulnerability to sea-level rise in the U.S. southeast. J Coast Res 12:327–338

Hereher M (2011) Mapping coastal erosion at the Nile Delta western promontory using Landsat imagery. Environ Earth Sci 64:1117–1125. https://doi.org/10.1007/s12665-011-0928-9

Kairu K, Nyandwi N (2000) Guidelines for the study of shoreline change in the western Indian Ocean region, IOC manuals and guides No. 40, UNESCO, p 55

Mujabar PS, Chandrasekar N (2013) Shoreline change analysis along the coast between Kanyakumari and Tuticorin of India using remote sensing and GIS. Arab J Geosci 6:647–664

Nandi S, Ghosh M, Kundu A, Dutta D, Baksi M (2016) Shoreline shifting and its prediction using remote sensing and GIS Techniques: a case study of Sagar Island, West Bengal (India). J Coast Conserv 20:61–80

Sestini G (1992) Implications of climatic changes for the Nile delta. In: Jeftic L, Milliman JD, Sestini G (eds) Climatic change and the Mediterranean. Edward Arnold, New York, p 5355601

Thieler ER, Hammar-Klose ES (1999) National assessment of coastal vulnerability to sea-level rise, US Atlantic Coast: US Geological survey, Open-file report, pp 99–593

UNDP (2009) Adaptation to climate change in the Nile Delta through integrated coastal zone management. United Nations Development Programme, New York, Project Document 3748, p 124

UNESCO/UNDP (1978) Coastal protection studies. Final Technical Report, Paris 1, pp 155

White K, El Asmar H (1999) Monitoring changing position of coastlines using Thematic Mapper imagery, an example from the Nile Delta. Geomorphology 29:93–105